Strain Measurement

Maria Teresa Restivo,
Fernando Gomes de Almeida,
Maria de Fátima Chouzal

Strain Measurement

**Measurement of Physical and Chemical Quantities Series
Vol. 1**

International Frequency Sensor Association Publishing

Maria Teresa Restivo,
Fernando Gomes de Almeida,
Maria de Fátima Chouzal
Strain Measurement
Measurement of Physical and Chemical Quantities Series, Vol. 1

Illustrations: Leonor Zamith
Programming of multimedia content: GATIUP, Universidade do Porto

ISBN-10: 84-616-0067-3
ISBN-13: 978-84-616-0067-0
BN-20120724-XX
BIC: TJF

To the memory of António Restivo and to his grand children
Constança, António and Luzia.

Maria Teresa Restivo

To my wife and daughter, Angela and Penelope.

Fernando Gomes de Almeida

To my parents.

Maria de Fátima Chouzal

Contents

Foreword **9**

Preface **11**

Chapter 1. Elementary Concepts of Stress and Strain **13**

1.1. Introduction *13*

1.2. Stress and Strain *14*

1.3. Linear Elasticity and Hooke's Law *15*

1.4. Shearing Stress and Angular Distortion *16*

1.5. Uniaxial Stress State *19*

1.6. Biaxial Stress State *21*

1.7. Plane Stress State *23*

1.8. Experimental Methods for Strain Measurement *27*

Chapter 2. Resistance Strain Gauges: Basic Concepts **29**

2.1. Working Principle *29*

2.2. Some Characteristics of the Resistance Strain Gauge *31*

2.3. Temperature Effect: Self-compensated Strain Gauges *36*

2.4. Strain Gauge Selection *38*

Chapter 3. Resistance Strain Gauges: Instrumentation and Techniques 43

3.1. Integration in a Wheatstone Bridge *43*

3.2. Lead Wire Effects *45*

3.3. Compensation of Temperature Effects *48*
3.3.1. Self-temperature-compensated Strain Gauges 48
3.3.2. Temperature Compensation with Half Bridge Configuration 53
3.3.3. Temperature Compensation with Full Bridge Configuration 55

3.4. Power Supply.. *56*

3.5. Signal Conditioning / Amplifiers *60*

Chapter 4. Fibre Optic Strain Gauges **65**

4.1. Introduction *65*

4.2. Interferometric Strain Sensing *67*
    4.2.1. Interferometric Strain Sensitivity 69

4.3. Strain Sensing with Fibre Bragg Gratings *71*
    4.3.1. Introduction to Fibre Bragg Gratings 71
    4.3.2. Strain Sensing 73
    4.3.3. Cross-sensitivity to Temperature 75

4.4. Impact of Strain Sensing with Fibre Bragg Gratings *76*

Chapter 5. Uncertainty Evaluation in Stress Measurement **79**

5.1. Basic Concepts *79*

5.2. Evaluation of Measurement Uncertainty: the Modelling Approach *81*

5.3. Uncertainty Evaluation Using the Modelling Approach *81*
    5.3.1. How to Estimate the Standard Uncertainty of Type A 83
    5.3.2. How to Estimate the Standard Uncertainty of Type B 83
    5.3.3. Law of Uncertainty Propagation 84

5.4. Evaluation of Uncertainty Associated to Stress Measurement *85*

5.5. Final Remarks *90*

Bibliography **93**

Media Files List and Locations **97**

Index **101**

Foreword

This book deals with measurement of stresses and strains in mechanical and structural components. This topic is related to such diverse disciplines as physical and mechanical sciences, engineering (mechanical, aeronautical, civil, automotive, nuclear, etc.), materials, electronics, medicine and biology, and uses experimental methodologies to test and evaluate the behaviour and performance of all kinds of materials, structures and mechanical systems.

During the last few decades the development of computer based techniques, as well as laser-optics methods, nanotechnologies and nanomaterials, among many other technological advances, added new dimension and perspectives to experimental techniques for testing, measurement and all related instrumentation.

The different subjects exposed in this book are presented in a very simple and easy sequence, which makes it most adequate for engineering students, technicians and professionals, as well as for other users interested in mechanical measurements and related instrumentation.

Joaquim Silva Gomes
Professor of Solid Mechanics and Experimental Mechanics

Preface

The present work, entitled "Strain Measurement", is the first volume in the book series "Measurement of Physical and Chemical Quantities". Two new titles, on temperature and on displacement measurement, will be published by the authors in the nearest future.

This first edition is devoted to strain measurement considering its relevance in the engineering field. Invited authors are also contributing to specific topics with valuable perspectives.

The text starts by introducing the elementary concepts of stress and strain state of a body. Next, several experimental extensometry measurement techniques are briefly introduced highlighting and covering the fundamental concepts of the mostly universal one: the electric resistance extensometry using electrical strain gauges. Basic instrumentation theory and techniques associated with the use of strain gauges are presented. Looking forward into modern advanced techniques optical fibre based extensometry is also covered. Taking into account the importance of the evaluation of measurement uncertainties the publication ends with the uncertainty estimation on the measurement of mechanical stress.

The different chapters include several multimedia components such as animations, simulations and video clips.

Chapter 1, *Elementary concepts of stress and strain*, presents fundamental concepts of the states of stress and strain. The understanding of these concepts is fundamental for the theme under study and aims at the accurate experimental evaluation based in the measurement of stress and strain states in loaded bodies (parts, structures...). A non exhaustive summary of some measurement techniques is also presented, emphasizing resistance extensometry, the most universal method.

Chapter 2, *Strain gauges: basic concepts*, conveys the basic concepts on strain gauges, namely their working principle and the most relevant characteristics. The meaning of self-compensated strain gauges is explained. Parameters that must be taken into account for the selection of a strain gauge are also referenced and some helpful guidelines are provided.

Chapter 3, *Resistance extensometry: instrumentation and associated techniques*, deals with a set of fundamental concepts on instrumentation, signal conditioning and various techniques that are instrumental to minimize the main causes of measurement error that are inherent to the use of strain gauges. Along the chapter the use of a few simple numerical calculations *"A few numbers!"* are applied in order to clarify some notions.

Chapter 4, *Strain measurement using optical fibres*, is written by the invited authors José Luís Santos, L.A. Ferreira and F.M. Araújo.

This chapter describes the use of the optical fibre as a sensing element which is being increasingly explored for measurement of physical quantities of which strain is a remarkable example. This chapter describes the working principle of optical fibre sensors and their application to strain measurement. Two sensorial structures are reviewed with detail, the Mach-Zehnder interferometer and the fibre Bragg grating, that may be considered as the optical equivalent of the electric resistance strain gauge.

Chapter 5, *Uncertainty estimation on the measurement of mechanical stress, is shared between Maria Teresa Restivo and* invited author Carlos Sousa.

Measurement is a fundamental activity in all engineering areas. But the result of a measurement action is not complete until the measured value is paired with information about its associated uncertainty. In this chapter the measurement of the flexural normal stress at the surface of a cantilever beam using resistance strain gauges is used as basis for uncertainty estimation.

The authors hope to have contributed with fundamental concepts and technical information on the topic of strain measurement. This contribution has been clearly enriched with the involvement of the invited authors.

The authors.

Chapter 1

Elementary Concepts of Stress and Strain

1.1. Introduction

The information about the stress state at one or more points of (mechanical) parts, such as gears and other components or structures, together with the mechanical material properties, is essential to predict or explain their mechanical response (or behaviour). This is the objective of stress analysis, a fundamental engineering tool for the construction of safe and reliable products and structures.

Alongside the development of analytical, numerical and numerical-analytical methods, experimental techniques have become extremely important, namely for confirming theoretical predictions of the real behaviour of experimental models or finished products. They are actually essential for some hybrid stress analysis methods.

Hence experimental stress analysis is nowadays a tool employed in various phases of the development and life of a product: from the testing of prototypes and finished products to assess their behaviour under overloading scenarios, up to the monitoring during their life cycle for fault or failure detection.

Experimental stress analysis may employ many different techniques which can be divided into two groups. There are those that investigate internal component behaviour through destructive methods or by the analysis of appropriate physical models, while others are based on external surface strain measurement.

1.2. Stress and Strain

The concepts of stress and strain may be illustrated in a simple way by considering the behaviour of a thin prismatic bar under axial tensile loading, Figure 1.1. The length of the bar, with constant cross section, will increase. A central portion of this bar, faraway from the bar ends where the loads F are applied and with an initial length l, undergoes an elongation, Δl, caused by the tensile forces, F.

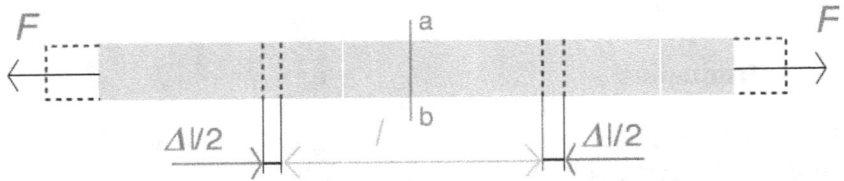

Figure 1.1. Prismatic bar under tensile loading.

Let us imagine the bar divided into two parts at section ab. It is now possible to consider one of them as a free body, the applied load F being balanced by forces that represent the action of the removed part, Figure 1.2. The intensity of those forces has an essentially uniform distribution over section ab. The force per unit area is called *stress* and is usually denoted by the Greek letter σ.

Figure 1.2. Stress at section ab.

Assuming the stress σ to be uniform, with a value at each point of section ab equal to its mean value, and taking into account the static equilibrium of the free body, the following expression can be derived relating stress, σ, cross sectional area, A, and applied load, F:

$$\sigma = \frac{F}{A} \qquad (1.1)$$

This equation shows that stress is force by unit area, measured in N/m^2 or Pa. When the loading elongates the bar, the stress is *tensile*, otherwise it is *compressive*.

The bar under tensile stress suffers an elongation Δl. The elongation per unit length or *unit elongation* is usually denoted by the Greek letter ε and assuming the elongation to be uniform along the bar length can be calculated as:

$$\varepsilon = \frac{\Delta l}{l} \qquad\qquad (1.2)$$

where l is the bar length. Note that unit elongation is dimensionless. It is nevertheless usual to employ a pseudo unit, the *micro strain*, to present numerical strain values (1 $\mu\varepsilon = 10^{-6}$ m/m).

When the bar is subject to tensile load there is *tensile deformation* that corresponds to an increase in length. If the bar is compressed there is *compressive deformation* and a reduction in length.

1.3. Linear Elasticity and Hooke's Law

Most structural materials display elastic behaviour under small applied loads. When an elastic material presents a linear relation between stress and strain it is called *linear elastic*. This is an important property of many solid materials, namely most metals, plastics, wood, ceramics and concrete. The linear relation between stress and strain may be expressed by the following equation:

$$\sigma = E \cdot \varepsilon, \qquad\qquad (1.3)$$

where E is a proportionality constant denominated *elasticity modulus* or *Young's modulus*, in tribute to the English scientist Thomas Young (1773-1829) who studied the elastic behaviour of bars. This equation is also known as *Hooke's law*, due to the work of another English scientist, Robert Hooke (1635-1703), who was the first to establish experimentally the linear relation between stress and strain.

When a bar is submitted to tensile loading the axial elongation is accompanied by lateral contraction: the bar width diminishes as its length increases. The ratio between the transversal, ε_t, and axial, ε_a, strains is constant within the elastic region and is called Poisson's ratio, usually denoted by the Greek letter v:

$$v = -\frac{\varepsilon_t}{\varepsilon_a} \qquad (1.4)$$

This coefficient is named after the French mathematician S.D. Poisson (1781-1840) who employed a molecular theory of materials to determine this relation. He found $v = 0.25$ for isotropic materials which present identical elastic properties in all directions. It can be proved that v reaches a maximum of 0.5 for materials that have zero volume change under tensile load. Experimentally it is shown that v is close to 0.3 for metals, near 0.5 for rubber and paraffin, almost zero for cork and around 0.2 for concrete.

1.4. Shearing Stress and Angular Distortion

Figure 1.3 illustrates a practical case where shearing stresses are present. The clevis yoke B_1 and the bar B_2 are connected by the screw B_3 that passes through the bar and clevis yoke. Under the action of load F the bar and clevis yoke tend to shear the screw at sections *ab* and *cd*.

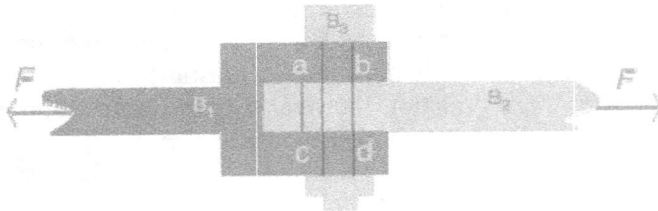

Figure 1.3. Connection subject to shearing forces.

The free body diagram of the portion of the screw bounded by sections *ab* and *cd* shows that the shearing forces V act on these sections (in this case $V = F/2$), Figure 1.4. The shearing force produces *shearing stress*, τ, on the screw cross-sectional area, A.

The exact distribution of this shearing stress is not easily determined. However a mean value can be obtained by dividing the shearing force V by the area A:

$$\tau = \frac{V}{A} \qquad\qquad (1.5)$$

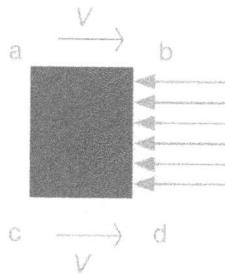

Figure 1.4. Shearing forces acting on the screw.

The tensile and compressive stresses previously referred act normally to the sections; therefore they are usually called *normal stresses*. On the other hand, the shearing stresses act tangentially to the surfaces; for this reason they are called *tangential stresses*. In both cases stress represents force per unit area, the main difference between the two types of stress being the direction.

Consider an elemental volume of material subject to shearing. Admit that a shearing stress, τ, acts on the top face, Figure 1.5. In the absence of any normal stresses acting on the cube its static equilibrium requires that a shearing stress of identical magnitude but opposite direction acts on the bottom face. These two stresses produce a moment that has to be balanced by an opposite moment originated by stresses acting on the vertical faces. Such stresses must also have equal magnitude τ, so that elemental equilibrium is achieved.

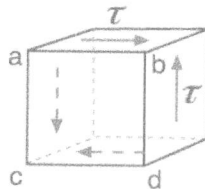

Figure 1.5. Elemental volume subject to shearing stresses.

In summary shearing stresses on a material element act in pairs with equal magnitude and opposite direction; the shearing stresses always exist in orthogonal planes with identical magnitude and either converging onto or diverging from the planes intersection.

When an element is only subject to shearing stress it is said to be in *pure shear*.

The deformation of the elemental volume caused by such shearing stress state is represented in Figure 1.6, where *abcd* is the frontal face. The length of the face edges does not change given that normal stresses are null in a state of pure shear. However the shearing stresses produce an *angular distortion* of the rectangular face into a parallelogram.

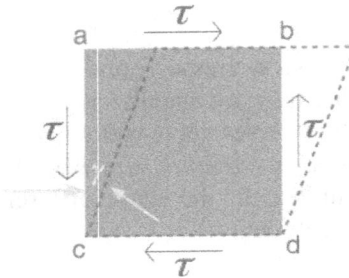

Figure 1.6. Angular distortion due to shearing stress.

The $\pi/2$ angle with vertex c, now measures ($\pi/2$ - γ), where γ is the angular distortion (or shearing strain) represented in the Figure 1.6. Conversely the $\pi/2$ angle with vertex *a* increases to ($\pi/2$ + γ) due to shear. The figure shows that the angular distortion is given by the ratio of the relative sliding between the top and bottom edges by the element height.

In a linear elastic material the angular distortion γ is proportional to shearing stress that causes it. So

$$\tau = G \cdot \gamma, \qquad (1.6)$$

where G is the *shear modulus*. As could be expected G and E are not independent; it can be shown that they are related by

$$G = \frac{E}{2 \cdot (1 + v)} \qquad (1.7)$$

1.5. Uniaxial Stress State

A uniaxial stress state is characterized by the existence of a single non-zero normal stress. If the x-axis is made to coincide with the stress direction, Figure 1.7, we have $\sigma_x \neq 0$, $\sigma_y = 0$ e $\tau_{xy} = 0$.

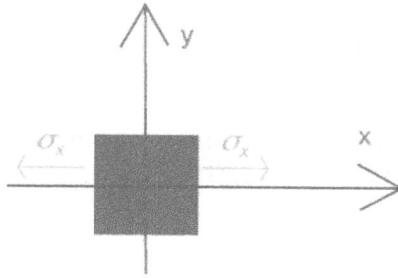

Figure 1.7. Uniaxial stress state.

The strains in the xy plane associated with this stress state are given by:

$$\varepsilon_x = \frac{1}{E} \cdot \sigma_x \qquad (1.8)$$

$$\varepsilon_y = -v \cdot \varepsilon_x = -\frac{v}{E} \cdot \sigma_x \qquad (1.9)$$

There is also strain along the z-axis given by:

$$\varepsilon_z = -v \cdot \varepsilon_x = -\frac{v}{E} \cdot \sigma_x \qquad (1.10)$$

Therefore the measurement of one of the strains ε_x or ε_y, or of both, may lead to the value of the σ_x stress.

It is interesting to observe what happens on a cross section not normal to the x-axis. Considering once more the prismatic bar subject to a

19

tensile load F, let us focus on a cross section whose normal makes an angle θ with the bar axis (x-axis), Figure 1.8.

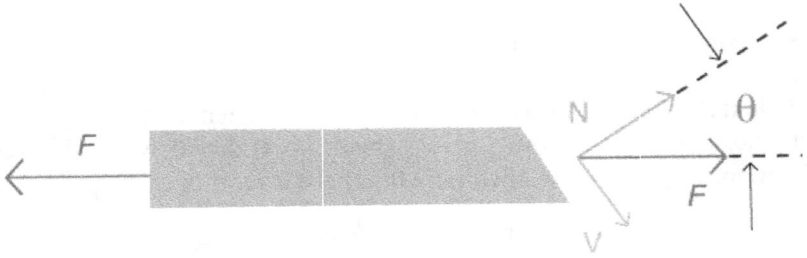

Figure 1.8. Forces acting on an oblique cross section.

The force F may be decomposed into two components - N, normal and V, tangential to that section:

$$N = F \cdot \cos\theta \qquad (1.11)$$

$$V = F \cdot \sin\theta \qquad (1.12)$$

As the oblique cross sectional area, A_θ, is $A/\cos\theta$, the stresses corresponding to N and V are, respectively:

$$\sigma_\theta = \frac{N}{A_\theta} = \frac{F}{A} \cdot \cos^2\theta = \sigma_x \cdot \cos^2\theta \qquad (1.13)$$

$$\tau_\theta = \frac{V}{A_\theta} = \frac{F}{A} \cdot \sin\theta \cdot \cos\theta = \sigma_x \cdot \sin\theta \cdot \cos\theta, \qquad (1.14)$$

where $\sigma_x = F/A$ is the stress on the cross section, normal to the bar axis (x stress).

From equations (1.13) and (1.14) it may be concluded that:

$$\sigma_{max} = \sigma_\theta(\theta = 0) = \sigma_x \qquad (1.15)$$

$$\tau_{max} = \tau_\theta(\theta = \pi/4) = \frac{\sigma_x}{2} \qquad (1.16)$$

1.6. Biaxial Stress State

A biaxial stress state is characterized by the existence of only two non-zero normal stresses, Figure 1.9. If the x-axis coincides with one of them and the y-axis with the other then $\sigma_x \neq 0$, $\sigma_y \neq 0$ and $\tau_{xy} = 0$.

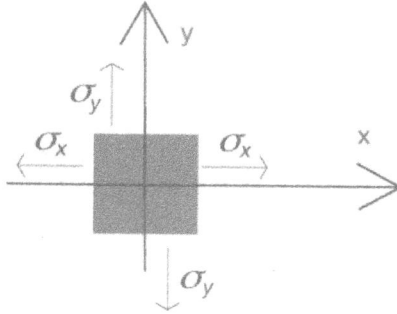

Figure 1.9. Biaxial stress state.

The strains in the *xy* plane associated with this stress state are given by:

$$\varepsilon_x = \frac{1}{E} \cdot \left(\sigma_x - v \cdot \sigma_y \right) \qquad (1.17)$$

$$\varepsilon_y = \frac{1}{E} \cdot \left(\sigma_y - v \cdot \sigma_x \right) \qquad (1.18)$$

From equations (1.17) and (1.18) can be derived the expressions that give the stress values from the strains:

$$\sigma_x = \frac{E}{1 - v^2} \cdot \left(\varepsilon_x + v\varepsilon_y \right) \qquad (1.19)$$

$$\sigma_y = \frac{E}{1 - v^2} \cdot \left(\varepsilon_y + v\varepsilon_x \right) \qquad (1.20)$$

Therefore the measurement of strains ε_x and ε_y leads to the value of stresses σ_x and σ_y. These stresses also originate strain along the *z*-axis, given by:

$$\varepsilon_z = -\frac{v}{E} \cdot \left(\sigma_x + \sigma_y \right) \qquad (1.21)$$

In order to determine the stresses acting on a surface whose normal makes an angle θ with the x-axis we may analyse the equilibrium of an elemental triangular prism obtained from the original elemental cube, Figure 1.10.

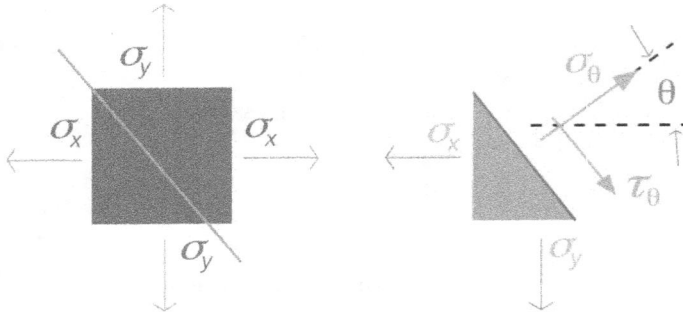

Figure 1.10. Stresses at an oblique section.

So, if A is the area of the x face (that is the face on which acts σ_x), the area of the y face is $A \cdot \tan \theta$ and the area of the oblique face $A \cdot \sec \theta$. The forces on faces x and y are, respectively, $\sigma_x A$ and $\sigma_y A \cdot \tan \theta$. Each one of these forces may be decomposed into two components, one normal and the other tangential to the oblique plane. The forces acting on the elemental prism may be projected onto these two directions. The projection onto the direction of σ_θ, the stress normal to the oblique face, leads to:

$$\sigma_\theta \cdot A \cdot \sec \theta = \sigma_x \cdot A \cdot \cos \theta + \sigma_y \cdot A \cdot \tan \theta \cdot \sin \theta \qquad (1.22)$$

or

$$\sigma_\theta = \sigma_x \cdot \cos^2 \theta + \sigma_y \cdot \sin^2 \theta \qquad (1.23)$$

The projection onto the direction of τ_θ, the stress tangent to the oblique face, gives:

$$\tau_\theta \cdot A \cdot \sec \theta = \sigma_x \cdot A \cdot \sin \theta - \sigma_y \cdot A \cdot \tan \theta \cdot \cos \theta \qquad (1.24)$$

hence,

$$\tau_\theta = (\sigma_x - \sigma_y) \cdot \sin\theta \cdot \cos\theta \qquad (1.25)$$

When the angle θ varies from 0 to $\pi/2$, the normal stress σ_θ varies from σ_x to σ_y. One of these stresses is therefore the maximum value of σ_θ and the other it's minimum. Such extreme values of the normal stress are called *principal stresses* and the corresponding directions *principal directions*.

The tangential or shearing stress, τ_θ, is zero when $\theta = 0$ and reaches the maximum value for $\theta = \pi/4$. Such value is:

$$\tau_{max} = \frac{\sigma_x - \sigma_y}{2} \qquad (1.26)$$

Therefore if the principal stresses are equal the shearing stress is zero in every direction.

1.7. Plane Stress State

The uniaxial and biaxial stress states are particular cases of a more general state, named plane stress. An elemental volume subject to plane stress may have normal and shearing stress on faces x and y; but stresses on face z are zero. The shearing stress on face x is denoted τ_{xy}, with the first index indicating the loaded face and the second the stress direction, Figure 1.11. Note that equilibrium considerations imply that $\tau_{yx} = \tau_{xy}$, as shown in *Section 1.4*.

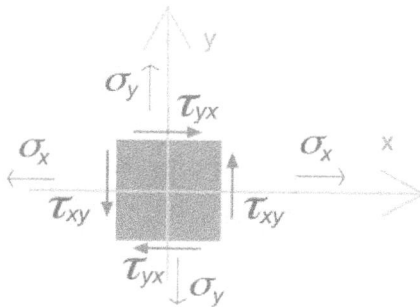

Figure 1.11. Plane stress.

The stresses on a plane at an angle θ with the x-axis may be determined following the same procedure used for the biaxial stress case. The normal stress, σ_θ, will then be:

$$\sigma_\theta = \sigma_x \cdot \cos^2\theta + \sigma_y \cdot \sin^2\theta + 2\tau_{xy} \cdot \sin\theta \cdot \cos\theta \quad (1.27)$$

and for the shearing stress, τ_θ, we get:

$$\tau_\theta = \left(\sigma_x - \sigma_y\right)\sin\theta \cdot \cos\theta + \tau_{xy} \cdot \left(\sin^2\theta - \cos^2\theta\right) \quad (1.28)$$

In this case the principal directions are not known beforehand. They may be determined by finding the zero of $d\sigma_\theta/d\theta$, which provides two values for θ, differing by $\pi/2$:

$$\tan 2\theta_p = \frac{2\tau_{xy}}{\sigma_x - \sigma_y} \quad (1.29)$$

The principal stresses, σ_1 and σ_2, are then obtained after some algebraic manipulations as:

$$\sigma_{1,2} = \frac{\sigma_x + \sigma_y}{2} \pm \sqrt{\left(\frac{\sigma_x - \sigma_y}{2}\right)^2 + \tau_{xy}^2} \quad (1.30)$$

In the cases previously considered of uniaxial or biaxial stress state, the *a priori* knowledge of the principal directions permitted the experimental determination of principal stresses from the measurement of strain along those directions. A different procedure has to be followed for the plane stress state, in which such directions are unknown. This procedure must also take into account that the sensors usually employed for strain measurement can only measure axial deformation and are unable to directly measure angular distortion.

The first and fundamental step for the definition of such procedure involves finding the relation between the strains ε_x, ε_y and γ_{yx} of the element and the axial strain, ε_θ, along a direction making an angle θ with the x-axis.

Figure 1.12 shows an elemental rectangle with sides dx and dy, whose diagonal coincides with the θ direction.

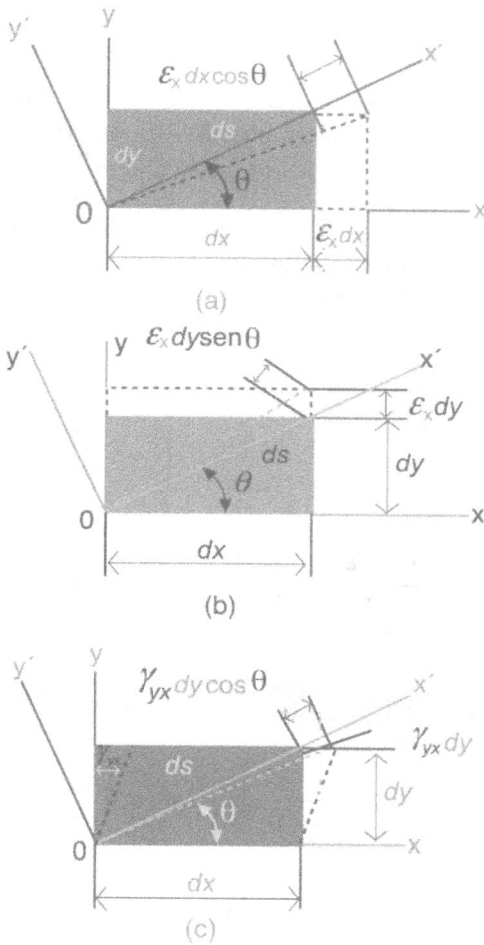

Figure 1.12. Deformation of an element due to: ε_x (a), ε_y (b) and γ_{yx} (c).

As a result of the ε_x, ε_y and γ_{yx} strains, the element elongates $\varepsilon_x\,dx$ along x and $\varepsilon_y\,dy$ along y, while the right angle xOy is reduced by γ_{yx}. Each one of these strains originates elongations of the diagonal whose values are, respectively, $\varepsilon_x\,dx.\cos\theta$, $\varepsilon_y\,dy.\sin\theta$ and $\gamma_{yx}\,dy.\cos\theta$. The total elongation of the diagonal is the sum of these three contributions. The corresponding strain, ε_θ, is obtained dividing the total elongation by the initial diagonal length, ds. Given that $dy/ds = \sin\theta$ and $dx/ds = \cos\theta$, one gets:

$$\varepsilon_\theta = \varepsilon_x \cdot \cos^2 \theta + \varepsilon_y \cdot \sin^2 \theta + \gamma_{yx} \cdot \sin \theta \cdot \cos \theta \qquad (1.31)$$

Measuring three strains, $\varepsilon_{\theta 1}$, $\varepsilon_{\theta 2}$ e $\varepsilon_{\theta 3}$, along three different directions, θ_1, θ_2 e θ_3, it is in this way possible to set up a system of three equations that leads to the values of ε_x, ε_y and γ_{yx}.

Following a procedure similar to that used above for stresses it is possible to compute from these three values the principal directions and the principal strains. One gets:

$$\tan 2\theta_p = \frac{\gamma_{yx}}{\varepsilon_x - \varepsilon_y} \qquad (1.32)$$

and

$$\varepsilon_{1,2} = \frac{\varepsilon_x + \varepsilon_y}{2} \pm \sqrt{\left(\frac{\varepsilon_x - \varepsilon_y}{2}\right)^2 + \left(\frac{\gamma_{yx}}{2}\right)^2} \qquad (1.33)$$

The following expressions, previously derived, provide the stress values from the measured strains:

$$\sigma_x = \frac{E}{1-v^2}\left(\varepsilon_x + v \cdot \varepsilon_y\right) \qquad (1.34)$$

$$\sigma_y = \frac{E}{1-v^2}\left(\varepsilon_y + v \cdot \varepsilon_x\right) \qquad (1.35)$$

$$\tau_{yx} = G \cdot \gamma_{yx} \qquad (1.36)$$

In particular the principal stresses are given from the principal strains by:

$$\sigma_1 = \frac{E}{1-v^2}\left(\varepsilon_1 + v\varepsilon_2\right) \qquad (1.37)$$

and

$$\sigma_2 = \frac{E}{1-v^2}\left(\varepsilon_2 + v\varepsilon_1\right) \qquad (1.38)$$

More complex stress states, such as the three-dimensional ones, can not be directly measured by means of current techniques. Nevertheless they may be obtained from the combination of measurements made in various faces of a specimen.

It should also be noted that in most cases the more important stresses (and strains) occur at the surface of the loaded bodies. Thus, the fact that the usual measurement methods only deal with surface strain is not normally an important restriction to the determination of critical stresses and strains that may lead to the failure of the components under observation.

1.8. Experimental Methods for Strain Measurement

The deformation along a line at the surface of a body is measurable with a mechanical strain gauge. Although this is the only direct method for strain measurement, it presents several inherent limitations, namely the absence of output in electrical or electronic form required for data storage and processing. Its use has therefore been discontinued.

Other methods, called indirect, are based on measurable physical phenomena produced by the strain experienced by a device bonded to the body or structure under study, such as:

- fragile rupture of a thin coating layer applied to the specimen surface – brittle coatings;

- variation of the vibration frequency of a metallic wire – vibrating-string strain gauges;

- fringe pattern obtained by superposing two grids with different pitch and orientation, one of which deforms with the specimen – *Moiré* (interferometry) method;

- variation of the electrical resistance of a conductor caused by axial strain – resistance strain gauges;

- bi-refringence induced by strain in special transparent plastic materials – photoelasticity;

- interference fringes resulting from the superposition of two holograms or laser gratings –speckle or laser interferometry.

At present there is intense activity in the field of fibre optic strain gauges, which are based in optical phenomena, as well as in that of thermal methods of stress analysis, such as SPATE (*Stress Pattern Analysis by Thermal Emissions*), TSA (*Thermographic Stress Analysis*) or TNDE (*Thermographic Nondestructive Evaluation*).

Contrarily to what might be expected, the resistance strain gauge technique, born in 1936, has neither disappeared nor suffered any drastic reduction in number and type of applications, in spite of the emergence of new alternative techniques. Its use continues to be widespread and recent developments are focused on the identification of new metallic alloys with more robust performance in hostile environments or demanding temperature conditions.

Moreover resistance strain gauges are perfectly adequate for measuring strains due to a large number of physical quantities of interest in many fields (such as strain, pressure, force, moment, acceleration, displacement,…). Figure 1.13 illustrates the application of resistance strain gauges in a mandibular study.

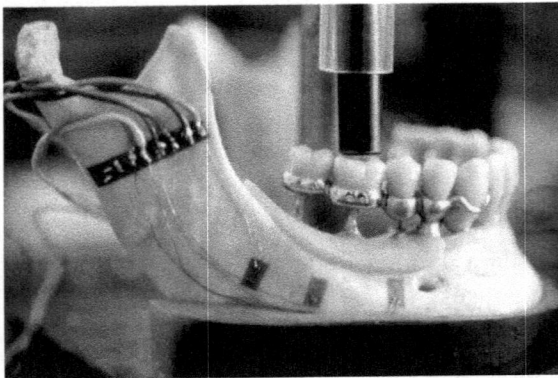

Figure 1.13. Human jaw under study using resistance strain gauges (kindly provided by LOME-FEUP).

In the discussion of strain measurement a particular emphasis will therefore be given to the description and application of resistance strain gauges.

Chapter 2

Resistance Strain Gauges: Basic Concepts

There are several methods for measuring strain. The most commonly used employs resistance strain gauges. Lord Kelvin first noted in 1856 that certain metallic conductors exhibited a change in electrical resistance when subject to strain.

Although a great variety of strain gauges are commercially available, the metallic resistance strain gauges of the bonded type (glued to the test part) are probably the most frequently used and will therefore deserve special attention in the following text.

2.1. Working Principle

The working principle of a resistance strain gauge is based in the fact that the electrical resistance of a material varies with its deformation. Consider a conductor or semiconductor element of length l, circular cross section of area A and resistivity ρ. The element resistance is a function of its geometry given by:

$$R = \rho \frac{l}{A} \qquad (2.1)$$

According to this expression the change in resistance, when the element is subject to a given tensile load, is due to the combined effect of the changes in length, in cross-sectional area and in material resistivity:

$$\frac{dR}{R} = \frac{dl}{l} - \frac{dA}{A} + \frac{d\rho}{\rho} \qquad (2.2)$$

When the strain gauge is well bonded to the surface of a test specimen they deform together. The strain gauge deformation along the longitudinal direction is identical to that which occurs on the specimen surface along the same direction.

$$\varepsilon_a = \frac{dl}{l} \qquad (2.3)$$

However the cross-sectional area also varies due to the Poisson effect, decreasing as the length increases, Figure 2.1. If the conductor has diameter D and Poisson ratio v, the transversal deformation will be given by:

$$\varepsilon_t = \frac{dD}{D} = -v \cdot \frac{dl}{l} \qquad (2.4)$$

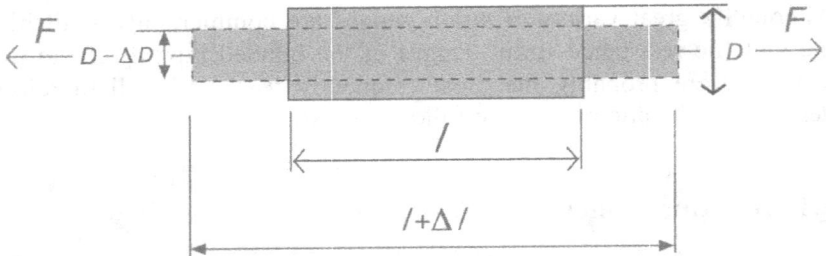

Figure 2.1. Dimensional changes and Poisson effect.

The cross-sectional area then varies according to:

$$\frac{dA}{A} = -2v\frac{dl}{l} + v^2\left(\frac{dl}{l}\right)^2 \qquad (2.5)$$

For small strains the second order term can be neglected, $\frac{dA}{A} \approx -2v\frac{dl}{l}$, and we finally get

$$\frac{dR}{R} = (1+2v)\frac{dl}{l} + \frac{d\rho}{\rho} \qquad (2.6)$$

30

This expression shows that the sensitivity of a material to strain, $S=\dfrac{dR/R}{dl/l}=(1+2v)+\dfrac{d\rho/\rho}{dl/l}$, is affected by changes in the length and cross-sectional area of the conductor or semiconductor, carried by the term $(1+2v)$, and by the piezoresistive effect of the material, $\dfrac{d\rho/\rho}{dl/l}$.

The sensitivity of metallic materials may vary from 2 to 4. In fact most of these materials have the Poisson ratio around 0.3 and the piezoresistive effect in the range 0.4 to 2.4. However, and especially in semiconductor materials, the piezoresistive effect can be much more significant, leading to substantially higher sensitivity values. This clear advantage is offset by a markedly nonlinear response and also a great sensitivity to temperature changes which consequently require more sophisticated signal conditioning circuits, making these materials less attractive.

2.2. Some Characteristics of the Resistance Strain Gauge

In its simplest form a strain gauge is composed by a very thin (conductor or semiconductor) wire deposited by a photolithography process in a grid pattern, as shown in Figure 2.2, with most of its length parallel to a given direction x.

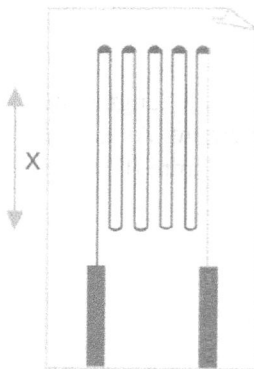

Figure 2.2. General layout of a resistance strain gauge of the bonded type.

Due to its fragility the grid is protected and supported by a very thin insulating film, the backing or carrier matrix. This has a double function. It facilitates the handling of the strain gauge during installation (particularly in the bonding operation). On the other hand it provides electrical insulation between the strain gauge and the device on which it is bonded. The support is bonded to the surface of the device whose strain is to be measured and it is responsible for transmitting this strain accurately to the grid. To this end the carrier matrix material must have special characteristics, namely in what concerns torsional stiffness, adhesion, insulation resistance, flexibility and ease of bonding, insensitivity to temperature, etc.

In this type of strain gauge the ratio between the relative change in resistance and the axial strain is called *Gauge Factor* (GF) and characterizes the sensitivity of the strain gauge to strain.

$$GF = \frac{\Delta R/R}{\Delta l/l} = \frac{\Delta R/R}{\varepsilon_a} \qquad (2.7)$$

This equation is related to that obtained for the sensitivity of a single conductor by substituting finite increments for the differentials.

However the existence of the end loops introduces some sensitivity to transversal strain. This undesirable effect may be substantially reduced by adopting special geometric configurations. Actually the end loops have larger cross sectional area than the longitudinal grid segments, contributing more to the nominal strain gauge resistance than to change in resistance under strain.

The layout of the conducting wire in a grid pattern provides strain gauges with a relatively small length and with a nominal resistance adequate for strain measurement. As defined in Figure 2.3 the length, l, may vary from 0.2 mm to 100 mm (0.008 in – 4 in).

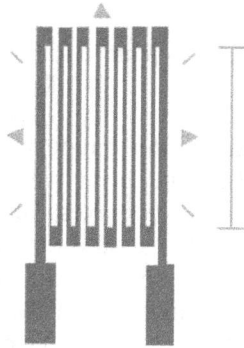

Figure 2.3. Definition of the strain gauge length.

Strain gauges with length below 3 mm tend to have inferior performance. Nevertheless such strain gauges should be used when the mounting space is very limited or when the objective is to measure localized strain in the vicinity of stress concentrations.

If there are no space constraints the use of larger strain gauges has some advantages. On the one hand larger dimensions (up to 13 mm length) are easier to produce and therefore cheaper, while handling and installation are easier. On the other hand the heat dissipation area increases (for a given resistance value) which is extremely important especially when the strain gauge is bonded to materials with poor heat transfer properties, such as plastics. Finally the use of a larger size strain gauge provides a better estimate of the average strain in heterogeneous materials.

Naturally very small or very large strain gauges are more expensive due to higher production cost. In the absence of particular conditions, such as those referred above, the choice of strain gauges with length from 3 mm to 6 mm is normally the most adequate. In this range strain gauges are available with a wide variety of characteristics which facilitates the selection taking into account other strain gauge parameters.

Strain gauges may be classified according to the number of grids (uniaxial, biaxial or multiaxial) and to the grid assembly (planar or stacked).

A uniaxial strain gauge with a single grid is normally limited to applications with uniaxial stress state and well known principal stress directions.

Biaxial stress states require strain gauge rosettes with two or three elements, respectively when the principal directions are known or unknown. In the first case the axes of the strain gauges are perpendicular, Figure 2.4 a), and should be lined up with the principal directions. In the second case the rosette may be installed without regard to orientation but generally one of the grids is aligned with the more significant axis of the test part. Two types of three-element rosettes are available: *rectangular* (with grids at 45°) or *delta* (grids at 60°), Figures 2.4 b) and c), respectively. The various rosette configurations are due to the great versatility of the photolithography process.

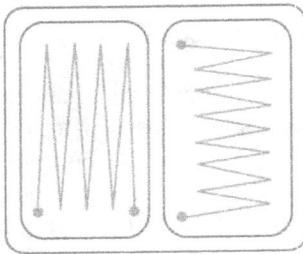

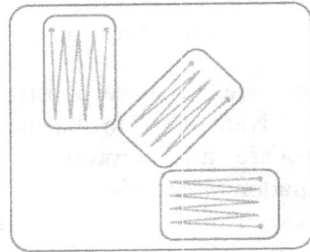

a) Tee or 90° rosette   b) Rectangular or 45° rosette

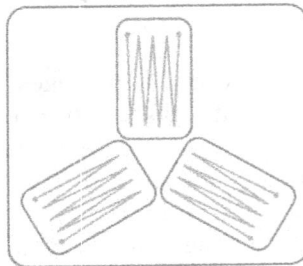

c) Delta or 60° rosette

Figure 2.4. Rosette grid orientation types.

In multiaxial rosettes the grid assembly may be *planar*, with all gauge elements lying in the same plane, or *stacked*, in which they are laid on top of each other - Figure 2.5.

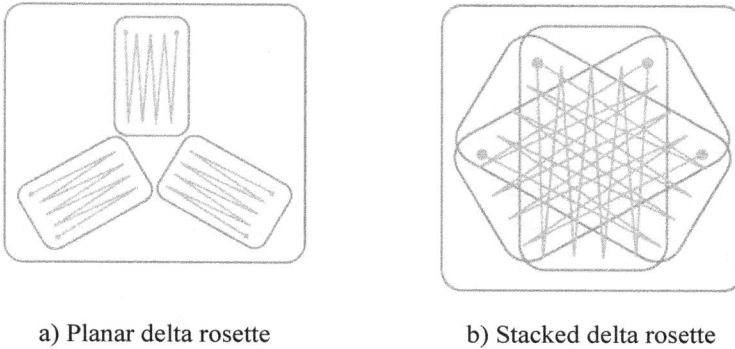

a) Planar delta rosette   b) Stacked delta rosette

Figure 2.5. Rosette construction types.

Planar rosettes, regardless of grid length, offer superior heat dissipation and generally provide more accurate static strain measurement since all grids lie on the surface of the test specimen. Stacked rosettes are indicated when there are space constraints or large strain gradients.

Unexpected as it may seem, only a small number of materials and alloys feature adequate properties for resistance strain gauge production. Among them the most frequently used are *constantan*, a copper-nickel alloy, and *karma*, a chromium-nickel alloy. Both have good sensitivity in a broad strain range and stability with changes in temperature. And their resistivity is sufficiently high to satisfy resistance requirements even with small grid length. These alloys are very versatile for the production of self-compensated strain gauges, whose thermal expansion coefficient is very closely adjusted to that of the test part material. *Constantan* is normally recommended for strain measurement at room temperature, whereas *karma* is used in extreme temperature conditions (either cryogenic or very high). Both alloys are adequate for static measurement.

For dynamic strain measurement it is more advantageous to use strain gauges produced with *isoelastic* materials which have superior fatigue resistance and also higher gauge factor that improves the signal to noise ratio.

The grid and the carrier materials are not selected independently nor combined arbitrarily. In fact the selection must be made from the available strain gauge *series*, each series corresponding to a specific combination of grid and carrier materials and consequently to distinct characteristics and preferential application areas.

Strain gauge electrical resistance has a typical nominal value of 120 Ω, but higher resistance values are also available (350 Ω, 500 Ω, 1000 Ω and 5000 Ω).

The 120 Ω strain gauges are the most frequently used due to their lower cost, especially in the case of those with very small dimensions. Actually it is more difficult from the technical standpoint to produce higher resistance strain gauges, because these require conductors with smaller cross sectional area which makes them much more fragile.

But they present some advantages if compatible with the instrumentation used. On one hand higher resistance leads to lower power dissipation (for a given voltage supply) and on the other hand the lead wire effects (wire resistance and its variation with change in temperature) will be less significant.

Typically the current applied to strain gauges is relatively small (from 2.5 mA to 40 mA) in order to avoid resistance changes due to self-heating and consequent thermal expansion.

2.3. Temperature Effect: Self-compensated Strain Gauges

Ideally strain gauges should present changes in resistance due exclusively to the strain of the surfaces on which they are bonded. However, in real applications several factors may also originate changes in resistance, giving rise to apparent strain in terms of measurement. Among them temperature is perhaps the most relevant. In fact the materials more adequate for strain gauge production are also sensitive to changes in temperature and the consequent changes in resistance may be of the order of those due to strain. Therefore when selecting a strain gauge its stability and sensitivity to temperature must also be taken into account.

The change in strain gauge resistance with temperature may be due to two distinct sets of factors. The first is related to the characteristics of the strain gauge itself: resistivity, dimensions and gauge factor.

The relative variation of strain gauge resistance due to changes in resistivity and in dimensions, $\dfrac{\Delta R_\rho}{R}$, is a function of the temperature coefficient of resistivity, α_ρ, and of the coefficient of thermal expansion of the strain gauge material, λ_e, given by:

$$\frac{\Delta R_\rho}{R} = \left(\alpha_\rho + \lambda_e\right)\cdot \Delta t = \alpha \cdot \Delta t, \qquad (2.8)$$

where α is the temperature coefficient of resistance of the strain gauge.

The contribution of gauge factor variation may be considered negligible for temperature changes below $50\,^\circ\text{C}$.

The second set of factors is related to the difference in coefficients of thermal expansion between the strain gauge and the material on whose surface it is glued.

The relative variation of strain gauge resistance due to this effect, $\dfrac{\Delta R_d}{R}$, is given by:

$$\frac{\Delta R_d}{R} = GF \cdot \left(\lambda_m - \lambda_e\right)\cdot \Delta t, \qquad (2.9)$$

where λ_m is the coefficient of thermal expansion of the test part material.

Combining both factor sets, the relative variation of strain gauge resistance with temperature, $\left(\dfrac{\Delta R}{R}\right)_{\Delta T}$, may be expressed as:

$$\left(\frac{\Delta R}{R}\right)_{\Delta T} = \left[(\lambda_m - \lambda_e)\cdot GF + \alpha\right]\cdot \Delta t \qquad (2.10)$$

Although desirable it is not possible to separate the apparent strain due to changes in temperature from the strain caused by load application. Nevertheless there are two methods that practically eliminate apparent strain. The first relies on the selection of a strain gauge material such that the factor $\left[(\lambda_m - \lambda_e) \cdot GF + \alpha\right]$ is zero. This is the basis of self-compensated strain gauges, commercially available. The second method compensates those effects by means of the integration in a Wheatstone bridge, as will be discussed in the next chapter.

Both *constantan* and *karma* may be subject to special metallurgy treatment for use in the production of self-compensated strain gauges which then have a thermal behaviour very close to that of some materials. In the strain gauge specification sheet the manufacturer indicates the material to which self-compensation is achieved.

Taking as an example the strain gauges of the *Vishay Micro-Measurements* Group, they include in their specification the so called '*STC number*' (*Self Temperature Compensation number*) that is approximately the coefficient of thermal expansion (expressed in ppm/°F) of the material on which the strain gauge will present minimum thermal output.

The '*STC number*' of the strain gauge to be used for a given static application should be as close as possible to the coefficient of thermal expansion of the test part material.

For purely dynamic strain measurements a more adequate strain gauge material is the D alloy (an isoelastic nickel-chromium alloy not subject to temperature compensation). In this case the two-digit STC number is replaced by the letters DY (for "dynamic").

2.4. Strain Gauge Selection

The correct and methodical selection of the characteristics and parameters of a strain gauge is fundamental for optimizing its performance under specific working and environmental conditions. This means increased measurement accuracy, greater ease of installation and minimum total cost.

The installation and the working characteristics of a strain gauge depend on the correct selection of the following parameters: material sensitivity to strain and carrier matrix, resistance, length, geometry and self-temperature compensation.

The strain gauge selection process involves some give and take but basically consists in the determination of a particular (and available) parameter combination that is the most adequate for the working conditions while satisfying all desirable requirements (in terms of accuracy, stability, insensitivity to temperature, ease of installation, etc.).

The strain gauge length and geometry are normally the first parameters to be defined, because they depend on the space available for installation and the expected stress state.

Both single grid and rosette strain gauges exist normally in several formats concerning grid width and solder tab location for compliance with the requirements of particular applications, as shown in Figure 2.6.

After the length and geometry selection the following step is the definition of the strain gauge series which determines the combination of strain gauge and carrier materials. They are not chosen independently and their combination is not arbitrary. The choice of the strain gauge material depends on the type of measurement (static or dynamic), on the working temperature, on the required accuracy and on the test part material. Manufacturers provide the characteristics of the existing series and, according to the strain gauge material selected, one has to choose the combination best suited for the intended application.

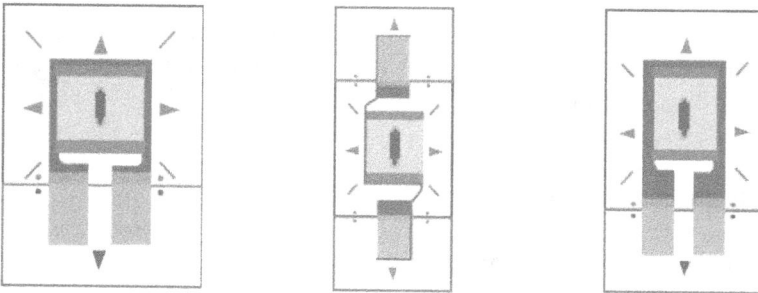

Figure 2.6. Single grid strain gauge models.

After the series selection one should check in the catalogue the availability of the desired length and geometry. Some adjustments may be required, such as opting for a slightly different geometry and length, or even choosing another series.

The choice of strain gauge resistance does not pose any major problem because practically all series offer the two standard values (120 Ω and 350 Ω).

The choice of the self-temperature compensation (STC) number is related to the type of test material and the required accuracy. Several numbers are provided for each strain gauge material and one should choose the one closest to the coefficient of thermal expansion of the test material. For example, the constantan A-alloy may have the following STC numbers: 00, 03, 05, 06, 09, 13, 15, 18, 30, 40, 50. If the test part material is carbon steel, whose thermal expansion coefficient is $6.7 \times 10^{-6}/$ °F, the strain gauge should have the STC number 06.

If we have instead an iron-nickel alloy, with a thermal expansion coefficient of $0.8 \times 10^{-6}/$ °F, the STC number of the strain gauge should be 00. In the case of pure tin, whose thermal expansion coefficient is $13 \times 10^{-6}/$ °F, the STC number should be 13.

As already mentioned the selection of all parameters involves some give and take and may require an iterative approach. Sometimes certain parameters have to be adjusted due to the lack of available strain gauges with the ideal characteristics for a given application.

A list is now given, in very general terms, of the main features to be taken into account for defining the various strain gauge parameters for a standard application.

Strain gauge length:

- Space available for installation;
- Ease of handling (longer strain gauges are easier to handle and to install);
- Heat dissipation (longer strain gauges are less sensitive to temperature effects);
- Type of test part material (homogeneous,...).

Strain gauge geometry:

- Stress state (uniaxial, biaxial, multiaxial);

- Installation space;
- Ease of installation;
- Heat dissipation.

Strain gauge series:

- Type of strain measurement (static, dynamic);
- Self-temperature compensation;
- Accuracy requirements;
- Ease of installation.

Strain gauge resistance:

- Heat dissipation;
- Lead wire effect;
- Instrumentation available.

STC number:

- Type of test part material;
- Temperature range;
- Accuracy requirement.

Chapter 3

Resistance Strain Gauges: Instrumentation and Techniques

3.1. Integration in a Wheatstone Bridge

The deformation-induced change in the resistance of a strain gauge must be converted into an electrical signal whose measurement will then enable the determination of that deformation.

Given that the relative change in the strain gauge resistance is very small, typically below 1 %, so is the change in the associated electrical signal, namely when compared with its nominal value (which is a function of the nominal value of the strain gauge resistance). For this reason the signal conditioning of the measurement system usually incorporates a Wheatstone bridge circuit.

Consider the Wheatstone bridge circuit of Figure 3.1, with one single resistance strain gauge R_e, bonded along the longitudinal direction of a test specimen subject to tensile load. The other bridge resistance values are identical to the nominal value of the strain gauge R_e. The bridge voltage supply value is V_0.

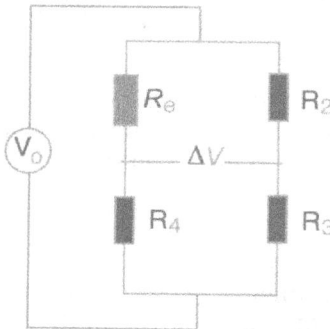

Figure 3.1. Wheatstone bridge with a resistance strain gauge.

As the change, ΔR_e, in the strain gauge resistance is very small when compared with its nominal value ($\Delta R_e \ll R_e$), the bridge unbalance value, ΔV, is given by:

$$\Delta V = V_0 \frac{\Delta R_e}{4 R_e}$$ (3.1)

The relative change in this unbalance is then related to the strain through the so called gauge factor (*GF*):

$$\frac{\Delta V}{V_0} = \frac{\Delta R_e}{4 R_e} = \frac{1}{4} \frac{\Delta l}{l} \quad GF = \frac{\varepsilon \, GF}{4}$$ (3.2)

If the Young modulus, E, of the test specimen material is known, the mechanical stress, σ, when the strain gauge region suffers an elastic strain of value ε, is:

$$\sigma = E \, \varepsilon$$ (3.3)

Taking into account the relations between the various quantities referred above, it is important to emphasize the advantage of integrating a resistance strain gauge into a Wheatstone bridge circuit. This will now be done with recourse to ...

A few numbers !

Let us analyse the example of Figure 3.2. An aluminium cantilever beam (E = 70 GPa = 70 × 10^9 N/m^2) of rectangular cross section is instrumented with a resistance strain gauge E_1 with 120 Ω nominal resistance and GF equal to 2, bonded on its upper face. If the vertical force F, applied at the beam free end, produces by bending a tensile stress of 1.05 MPa (1.05 × 10^6 N/m^2) in the zone where the strain gauge is bonded, its resistance undergoes a relative change of 0.003 % ! If a 10 mA current is supplied to the strain gauge, the voltage change at its terminals is 36 μV in 1.2 V.

Consider the same strain gauge now included in a Wheatstone bridge circuit, as illustrated in Figure 3.1. Admit that the other resistances are 120 Ω and that the bridge voltage supply is 2.4 V (so that the strain gauge current is around 10 mA, as in the previous case). The bridge unbalance value due to a 0.003 % increase of the strain gauge resistance is 18 µV around zero.

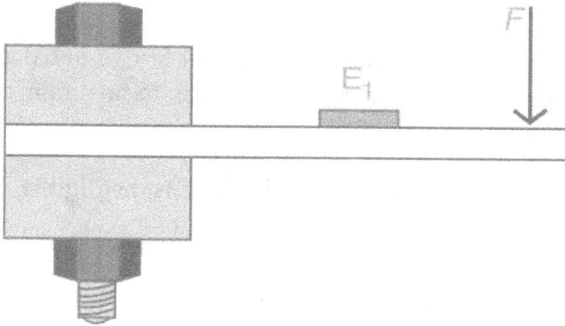

Figure 3.2. Aluminium cantilever beam instrumented with a resistance strain gauge.

In other words, now that the bridge unbalance signal is merely one half of that obtained previously, the measurement resolution has been significantly improved because the signal change is now around zero.

Thus the resolution can be expected to improve by a factor of at least 100 for ordinary measurement equipment and by 1000 for top quality equipment.

3.2. Lead Wire Effects

The resistance introduced by the lead wires has not been considered yet. Its value may however be significant in comparison with the small, deformation-induced, change in strain gauge resistance.

We shall now discuss the influence of lead wire resistance when an active strain gauge is connected to a bridge circuit.

Let consider a quarter bridge configuration. The other resistance values are identical to the nominal value of the strain gauge resistance. Assume that the change in strain gauge resistance due to deformation is very small, i.e. $\Delta R_e \leq 1 \% R_e$, and that the bridge is initially balanced (which may require the integration of an adjusting potentiometer, for example by inserting in Figure 3.1 a potentiometer R_p in series with R_4). Let us also admit that there are no temperature effects on the resistance of the lead wires.

The expressions that relate the measurement bridge unbalance with the change in the strain gauge resistance due to a strain, ε, are now presented for three distinct cases.

1) The resistance of the two lead wires is negligible, i.e. $R_L \cong 0$ ($R_p = 0$):

$$\Delta V = \frac{V_0}{4R} \Delta R = K_0 \Delta R \qquad (3.4)$$

2) The resistance of the two lead wires is no longer negligible and has the same value, R_L, for each ($R_p = 2 R_L$):

$$\Delta V = \frac{V_0}{4R + 8R_L} \Delta R = K_2 \Delta R \qquad (3.5)$$

3) The strain gauge layout uses the 3-wire method, as shown in Figure 3.3. The 3 wires and their resistances are identical ($R_{L1} = R_{L2} = R_{L3} = R_L$ and $R_p = 0$):

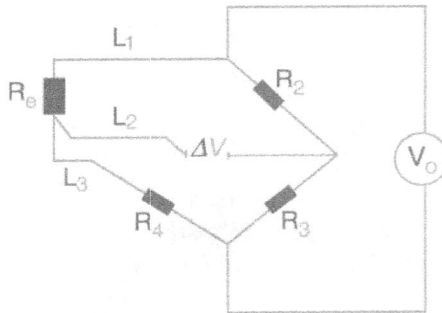

Figure 3.3. Strain gauge in a measurement bridge with a 3-wire configuration.

In this case the bridge unbalance is given by:

$$\Delta V = \frac{V_0}{4R + 4R_L} \Delta R = K_3 \Delta R \qquad (3.6)$$

In expressions (3.4), (3.5) and (3.6), the ratios of ΔV to ΔR, respectively, K_0, K_2 and K_3 correspond to the measurement bridge sensitivity for each case considered. They highlight very distinctly the varying influence of the lead wires on the sensitivity.

The benefits of the 3-wire method become more obvious by carefully comparing the three situations described above with the help of ...

A few numbers !

In the three cases above consider that the strain gauge nominal resistance (R_e) is 120 Ω, as well as the other bridge resistances. In all situations the bridge supply voltage is 2.5 V. In case 1 the lead wire resistance is considered negligible. In case 2 assume that the two lead wires have identical resistance R_L and that the same happens for the three wires in case 3.

Table 3.1 presents the values of K_0, K_2 and K_3 computed for a strain gauge change in resistance, ΔR_e, of 0.6 Ω and for two different values of lead wire resistance (corresponding to wires with small length). The variations of K_2 and K_3 relative to K_0 have also been determined.

Table 3.1. Lead wire resistance and bridge sensitivity.

R_e (Ω)	ΔR_e (Ω)	RL (Ω)	K_0 mV/(Ω)	K_2 mV/(Ω)	K_3 mV/(Ω)	$[(K_2-K_0)/K_0]$ × 100 %	$[(K_3-K_0)/K_0]$ × 100 %
120	0.6	0.1	5.208	5.200	5.204	- 0.17	- 0.08
120	0.6	0.5	5.208	5.165	5.187	-0.83	- 0.41

Taking as reference the ideal situation – case 1 – the following conclusions may be extracted from Table 3.1:

i)   The reduction of bridge sensitivity for cases 2 and 3 increases with the resistance of the lead wires. However the sensitivity is less penalized in case 3, by more or less half of the percentage obtained for case 2.

ii) Case 2 requires a potentiometer in the bridge to adjust the offset caused by the insertion of the strain gauge and the associated lead wire resistances.

iii) In theory case 3 can do without the adjusting potentiometer because the layout cancels the wire effect, regardless of their resistance or length, provided it is the same for all three. In fact, considering the bridge balance, L_1 and L_3 compensate each other since they belong to adjacent arms, while L_2 is connected to a high impedance measurement device which causes the wire effect to be negligible.

iv) It should be noted that if the change in lead wire resistance with temperature had been considered, as will be discussed in the next section, no compensation would have been possible in case 2, making the situation even worse in terms of bridge sensitivity. On the other hand compensation of such effect is possible in case 3.

In conclusion the 3-wire method has considerable advantages but nevertheless it does not eliminate the reduction of bridge sensitivity.

3.3. Compensation of Temperature Effects

3.3.1. Self-temperature-compensated Strain Gauges

In practical terms the change in resistance of a strain gauge due to deformation of the material onto which it is bonded is generally also affected by other sources. Among them temperature has undoubtedly the most disturbing effect. Temperature changes cause changes in the strain gauge resistance that are indissociable from those with mechanical origin and which are translated into apparent strain of thermal origin (ε_{at}). Such apparent strain may be significant in some cases.

Apparent strain must therefore be corrected or its causes minimized.

Using resistance strain gauge production technology it is possible to make the so called self-temperature-compensated strain gauges by drawing near the thermal expansion coefficients of both the strain gauge and the test specimen materials. The use of strain gauges self-temperature-compensated for the type of material under test minimizes the temperature effects within certain limits. In applications with high accuracy requirements the use of such strain gauges still demands additional corrections, as will be explained later.

Yet self-temperature-compensated strain gauges are not available for all types of materials. But we shall see that there are other ways to minimize this type of disturbance.

Translating into numbers the above comments, this effect is now illustrated by analysing ...

A few numbers!

The following examples use technical information kindly made available by the Micro Measurements Group (MM).

Consider a carbon steel bar on which is bonded a constantan strain gauge whose apparent thermal strain is plotted in Figure 3.4.

This strain gauge has been connected to a quarter bridge. The measurement bridge has been balanced at 24 °C room temperature in total absence of mechanical stress. But the test was performed at 34 °C temperature. There is a 10 °C difference between the bridge balance and the test operations which will be reflected in the measurement result as an apparent strain of thermal origin.

Taking into account the limitations imposed by the resolution of Figure 3.4 the apparent strain of thermal origin at 34 °C is approximately -100 με. At 100 °C the value would be around -1000 με!

Now suppose that we employ a self-compensated constantan A-alloy strain gauge. These strain gauges are available with various STC-numbers (according to MM): 00, 03, 05, 06, 09, 13, 15, 18, 30, 40 and 50. Given that the coefficient of thermal expansion of carbon steel is $6.7\times10^{-6}/$ °F $(12.1\times10^{-6}/$ °C), the more adequate strain gauge STC-number for this application is '06'.

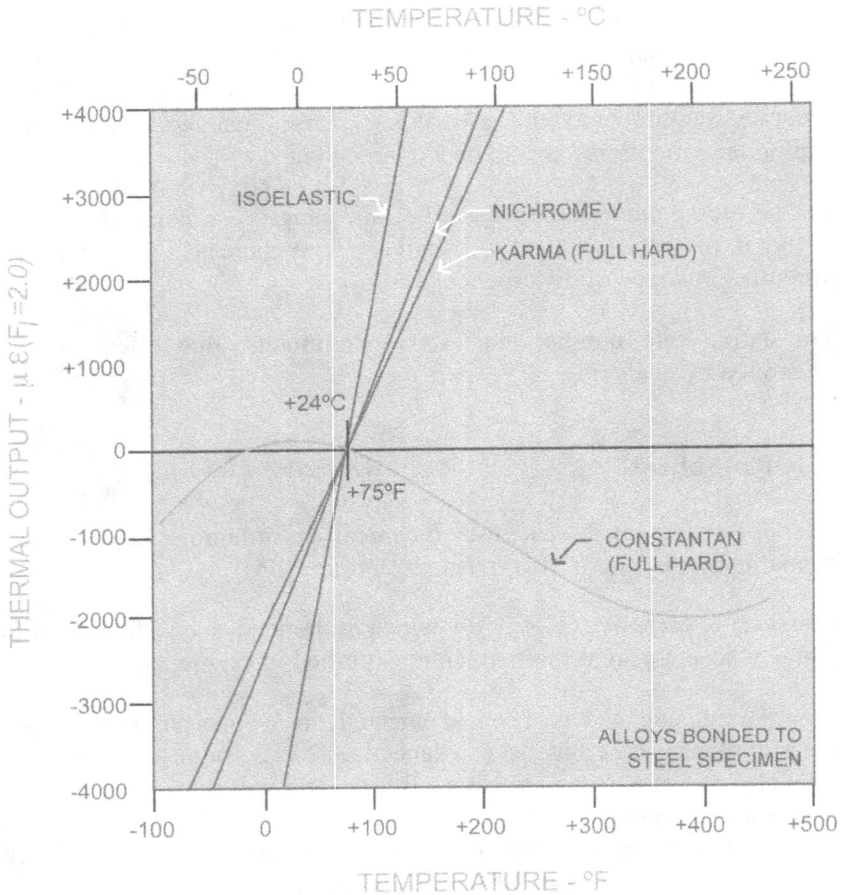

Figure 3.4. Apparent strain of thermal origin for strain gauges bonded on steel.

Consider, for example, the strain gauges with reference CEA-06-187UW-120 from the lot to which the manufacturer provides the following polynomial for apparent strain of thermal origin (in μm/m, t $^\circ$C):

$$\varepsilon_{at} = -3.29 \times 10 + 2.79 \times t - 6.56 \times 10^{-2} \times t^2 + 3.54 \times 10^{-4} \times t^3 - 4.29 \times 10^{-7} \times t^4$$

Using this polynomial one concludes that at 34 $^\circ$C the apparent strain of thermal origin is merely -0.5 $\mu\varepsilon$! Table 3.2 displays some values of ε_{at}, calculated with the polynomial for various temperatures.

Table 3.2. Apparent strain of thermal origin for CEA-06-187UW-120.

Temperature (°C)	ε_{at} (µε)
24	1.0
34	-0.5
60	-30.8
100	-98.8
150	-112.8
200	46.7
-50	-383.3
-60	-518.5

It has been mentioned that the polynomial supplied by the manufacturer corresponds to the strain gauges of a given lot. In fact the same type of strain gauge from a different lot will have, in principle, a different polynomial. Figure 3.5 shows the curve of apparent strain vs. temperature, as well as the ε_{at} polynomial for the CEA-06-187UW-120 strain gauges of a particular lot. This type of technical information is provided for any strain gauge pack of this manufacturer.

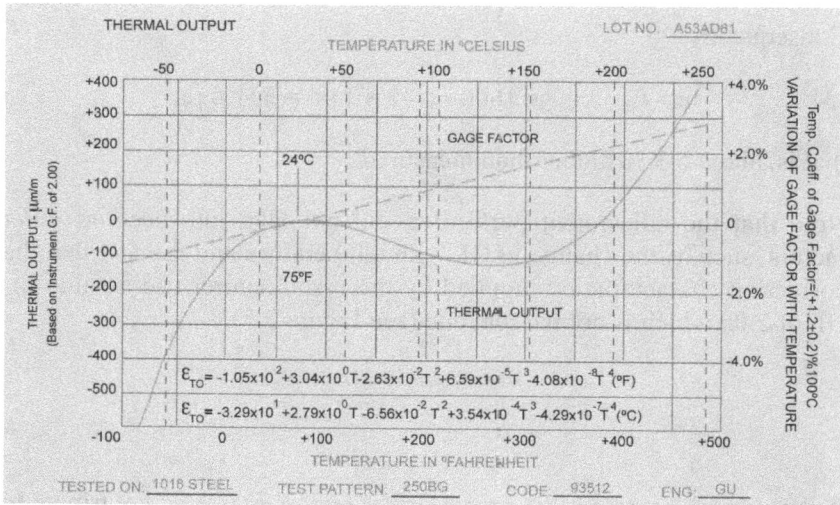

Figure 3.5. Apparent strain of thermal origin for the CEA-06-187UW-120 strain gauges of the lot referenced in the top right corner.

As shown by Table 3.2 the apparent strain of thermal origin of a self-temperature compensated strain gauge may be significant at both high and low temperatures. Therefore additional corrections may be necessary with this type of strain gauges, depending on the working temperature and also the accuracy required.

Let us once again look at ...

A few numbers !

Admit that the accuracy requirement demands corrections, even using a self-temperature compensated strain gauge for carbon steel. The measurement system has been balanced at 24 °C room temperature and without mechanical stress on the test specimen. Using the curve and the polynomial supplied with the strain gauge the apparent strain at 24 °C is +1.03 με. Suppose however that the test temperature is 60 °C. At this temperature the measurement result is 600 με. The corresponding apparent strain of thermal origin is now $\varepsilon_{at} = -30.8$ με. Therefore the partially corrected value, ε_{atc}, is:

$$\varepsilon_{atc} = \varepsilon_{at}(60\ ^\circ C) - \varepsilon_{at}(24\ ^\circ C) = (-30.8 - 1.0)\ \mu\varepsilon = -31.8\ \mu\varepsilon$$

Consequently,

$$\varepsilon = \varepsilon_{med} - \varepsilon_{atc} = (600 - (-31.8))\ \mu\varepsilon = 631.8\ \mu\varepsilon$$

that is, about 5.3 % higher than measured.

Note that the adjustment performed did not take into account other factors, such as the change in GF with temperature and the fact that the correction characteristics supplied by the manufacturer use a value for GF of 2.00 which is not the real one (see Figure 3.5).

It should also be noted that several other factors may also contribute to temperature related errors which are not always accounted for when using a self-temperature compensated strain gauge. The material properties and the shape of the test specimen, the strain gauge geometry (rosette, ...), the change in GF and in transversal sensitivity with temperature, the strain gauge lot and series, the changes of other resistances in the electrical circuit and the strain gauge installation procedure may also be error sources in strain measurement.

It is not always possible to employ self-temperature compensated strain gauges to minimize the temperature effect. In such cases the minimization of that effect may be achieved by conveniently integrating the strain gauges in a Wheatstone bridge. Actually this method may be always employed and is analysed in the following sections.

3.3.2. Temperature Compensation with Half Bridge Configuration

The error due to apparent strain of thermal origin may be considerably reduced with recourse to the electrical circuit properties of the Wheatstone bridge.

Consider a half bridge configuration with two identical strain gauges in adjacent bridge arms as shown in Figure 3.6.

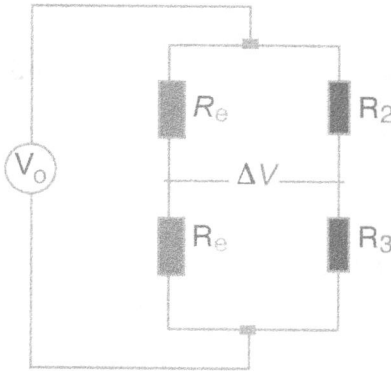

Figure 3.6. Half bridge configuration.

Two distinct possibilities are available:

i)   One of the two strain gauges (called active) is bonded to the test specimen while the other (named dummy) is glued to a specimen of the same material that is not subject to any mechanical stress but only to the working temperature effect. The two specimens should be as close to each other as possible;

ii)   Both strain gauges are active and bonded to the test specimen but in zones subject to symmetrical strain.

The first case relies on the use of a dummy strain gauge whose series and characteristics are in principle identical to the active one. If both are bonded to specimens of the same material and subject to identical environmental conditions, then the temperature effects on the change in resistance of both strain gauges are of the same order of magnitude and of the same type. The sensor integration in adjacent bridge arms leads naturally and theoretically to the elimination of the temperature effect contribution to the measurement bridge signal. In this case the lead wires for both strain gauges should have the same material, length and working temperature. If this is ensured then the temperature changes in both will induce identical changes in the lead wire resistances which will be cancelled by the bridge circuit intrinsic properties already described.

This process would fully eliminate temperature effects if all the listed conditions were fulfilled. However this method is sometimes difficult to implement due either to space constraints for the installation of the second unstressed specimen with the dummy strain gauge, or to different temperature conditions (especially when there are temperature gradients or transients around the test specimen). The same problem happens in what concerns the lead wires. These aspects are particularly important when the measurements are carried out in extreme temperature conditions: at high temperature or at cryogenic temperature.

In the second case the use of two active strain gauges overcomes some of the problems. In this technique both strain gauges are bonded to the test specimen. They and their lead wires are therefore subject to the same change in resistance due to changes in temperature, which minimizes the bridge unbalance due to temperature effects. In this way the unbalance measured will be fundamentally due the change in resistance due to strain. Once more we speak about minimizing and not eliminating the effect because the strain gauges are only in theory identical. In this second case the same precautions should apply to the lead wires. Figure 3.7 a) and b) illustrates two examples with two active strain gauges.

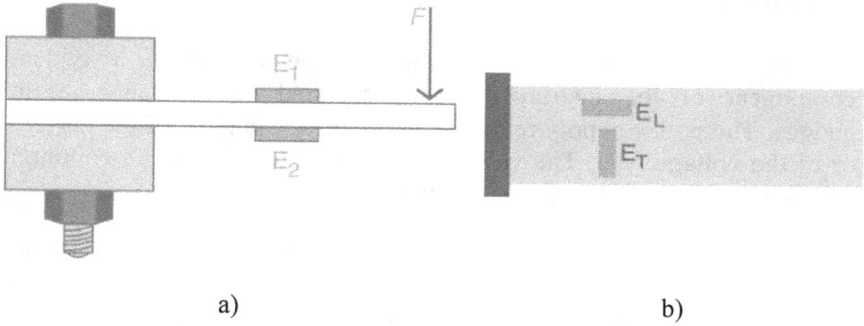

a)   b)

Figure 3.7. Beam instrumented with two strain gauges.

3.3.3. Temperature Compensation with Full Bridge Configuration

Whenever the possibility exists of using a full bridge circuit configuration the increase in the measurement system sensitivity adds to the advantages already described in the previous section. Such is the case depicted in Figure 3.8. In that particular layout all four strain gauges contribute to the elimination in theory (minimization in real terms) of the temperature effect (that originates changes in resistance that are very close in each gauge).

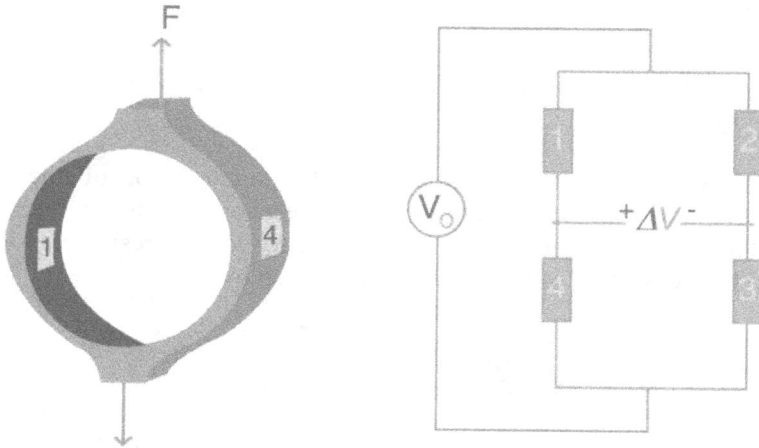

Figure 3.8. Test piece instrumented with 4 strain gauges and corresponding full bridge layout.

3.4. Power Supply

As mentioned before a Wheatstone bridge circuit is a typical component of the instrumentation associated to resistance strain gauges. The power supply required by the bridge may be of the current or of the voltage type. The voltage power supply is the more common and may be direct (DC) or alternating (AC).

An AC voltage power supply requires a signal demodulation circuit, which may limit dynamic measurements (band-pass $\approx$ 10% the frequency of the carrier wave) and introduce parasitic capacitance. It is therefore convenient to use a DC voltage power supply with good stability characteristics.

In order to get accurate results with a generic strain measurement system, like that of Figure 3.9, it is essential to maximize the sensitivity of the input block, thereby providing the system with a good signal to noise ratio.

Figure 3.9. Strain measurement system.

In what follows various parameters that may influence the input block sensitivity will be analysed, starting with the Wheatstone bridge power supply.

Consider in the present case that the input block of the strain measurement system comprises a Wheatstone bridge circuit with voltage supply V_0, one active strain gauge $R_e=R_1$, and resistances R_2, R_3, R_4. The sensitivity of the input block is given by:

$$\frac{\Delta V}{\varepsilon} = V_0 \, GF \, \frac{R_2 R_3}{(R_2 + R_3)^2} \qquad (3.7)$$

The sensitivity value cannot be significantly changed by the gauge factor (GF), which in the case of metallic resistance strain gauges has a limited range, as we already know. It is nevertheless possible to operate upon V_0 or upon the factor that congregates the bridge resistances, that is an intrinsic characteristic of the bridge circuit.

It is not however possible to analyse in a simplistic manner the sensitivity increase with the increase of V_0, which might lead to unrealistic values for the power supply. A more detailed analysis is therefore required.

The voltage supply V_0 may be expressed as:

$$V_0 = i_e (R_e + R_4) = i_e R_e (\frac{R_2 + R_3}{R_2})$$
(3.8)

where i_e represents the electric current flowing through the strain gauge.

Substituting V_0 in equation (3.7) we get:

$$\frac{\Delta V}{\varepsilon} = \sqrt{P R_e} \; GF \; (\frac{R_3}{R_2 + R_3}),$$
(3.9)

where P is the electric power dissipated by Joule effect on the strain gauge.

The electric power dissipated by a strain gauge depends on various factors. Being limited by the maximum current intensity allowed for the strain gauge, it also depends on the thermal conductivity and dissipation capacity of the test piece on which the strain gauge is bonded. Taking into account all these factors and introducing for each test piece type the concept of dissipated power density ($p_D = P/A$), where A is the area of the strain gauge grid, equation (3.9) may be rewritten as:

$$\frac{\Delta V}{\varepsilon} = \sqrt{A p_D R_e} \; GF \; (\frac{R_3}{R_2 + R_3})$$
(3.10)

Table 3.3 presents, as an example, recommended value ranges for p_D, in the case of 350 Ω, MM strain gauges for various test piece materials and characteristic.

Table 3.3. Dissipated power density.

Power density (W/mm^2)	Test piece characteristics
0.008 - 0.016	Aluminium, copper, with thick cross section
0.003 - 0.008	Steel, with thick cross section
0.0015 - 0.003	Steel, with thin cross section
0.0003 - 0.0008	Fibre glass, glass, ceramics
0.00003 - 0.00008	Plastics

The set of parameters that can be adjusted to increase the system sensitivity may be identified from equation (3.10):

- Parameters that depend on the strain gauge characteristics and/or of their association with the material on which it is bonded: the nominal value of the strain gauge resistance (for which there are limitations, as we saw), the area A of the strain gauge grid (within the dimensional constraints of each application), the gauge factor GF (which does not change much) and the power dissipation density p_D, which is a function of the power the strain gauge is able to dissipate when bonded on a given type of material.

- One parameter that depends on the Wheatstone bridge circuit design and is a function of the resistance values: $[R_3/(R_2 + R_3)]$. This factor takes the value 0.5 when all the resistances are equal to the nominal value of the strain gauge resistance. Raising it above 0.75 to 0.8 leads to unrealistic voltage supply values.

Therefore, in order to define the Wheatstone bridge voltage supply value, V_0, the strain gauge manufacturer data should be taken into account. Depending on the way this data is supplied we have two alternative starting points:

- The characteristics defined for each strain gauge when only the upper limit for the bridge voltage supply is defined;

or

- The manufacturer characteristics of each strain gauge type when these disclose A, R_e, GF and p_D and their conjugation with the bridge design (R_2, R_3 and R_4, with $R_4 R_2 = R_3 R_e$).

In the latter case it is convenient to rewrite V_0 as:

$$V_0 = \sqrt{A\, p_D\, R_e}\ \ (\frac{R_2 + R_3}{R_2})\qquad(3.11)$$

It should be noted that this analysis assumed the existence of only one active strain gauge. However there may be two or even four active strain gauges. In such cases the expression (3.11) would have to be adapted to the new situations.

We will now clarify these questions using values that the parameters may have in reality and comparing the results computed for the measurement system sensitivity and for the admissible power supply value in various situations with the help of ...

A few numbers !

Let us consider six distinct cases, all of them with a Wheatstone bridge with a single active strain gauge, always of the same type. Table 3.4 fully describes each case, presenting in the two final rows the computed values of the system sensitivity, $\Delta V/\varepsilon$, and the associated power supply value, V_0.

Table 3.4. System sensitivity and associated bridge power supply value.

	Case (i)	Case (ii)	Case (iii)	Case (iv)	Case (v)	Case (vi)
Nominal resistance (Ω)	350	350	350	350	350	350
GF	2	2	2	2	2	2
Grid area (mm^2)	49	49	9	60	49	49
p_D(W/ mm^2)	0.016 (*)	0.0015 (**)	0.0015	0.016	0.016	0.016
R_2 (Ω)	350	350	350	350	350	350
R_3 (Ω)	350	350	350	350	870	3150
R_4 (Ω)	350	350	350	350	870	3150
$R_2/(R_2 + R_3)$	0.5	0.5	0.5	0.5	0.7	0.9
$\Delta V/\varepsilon$ (V/$\mu\varepsilon$)computed	16.6E-6	5.07E-6	2.17E-6	18.3E-6	23.6E-6	29.8E-6
V_0 (V)computed	33.1	10.1	4.35	36.7	57.3	165.6

(*) copper test piece with thick cross section, (**) steel test piece with thin cross section

The analysis of the table leads to the following conclusions:

- The increase of the area of the strain gauge grid provides better thermal dissipation conditions, for a given current value. A higher power supply value, V_0, is therefore possible and, consequently, the measurement system sensitivity improves, as shown in cases 1 and 4.

- Better thermal conduction and dissipation properties of the test piece on which the strain gauge is bonded, provide better thermal dissipation to the strain gauge. Consequently, it is possible to use a higher voltage supply which leads to better system values, as shown in cases 1 and 2.

- High values for the $R_2/(R_2+R_3)$ factor may lead to unrealistic power supply values. The unit able to supply the power for case 6 can hardly be expected to possess the stability characteristics required in such applications.

3.5. Signal Conditioning / Amplifiers

The signal conditioning of strain measurement systems integrating resistance strain gauges typically involves a (quarter, half or full) Wheatstone bridge circuit.

The Wheatstone bridge circuit has been treated so far as if it was possible in practice to set it up with precisely identical resistance values in order to have zero unbalance in the absence of gauge strain. In reality such identity cannot be achieved. Moreover, when there are several active strain gauges their nominal resistance values always differ slightly.

On the other hand the bonding procedure may introduce residual stresses that add to the difference in nominal resistance values (in the case of a 120 Ω nominal resistance strain gauge a 1 % increase may be observed after bonding!).

In addition the strain gauge lead wires may be slightly different which also contributes, even minutely, to the lack of uniformity of the four Wheatstone bridge arm resistances.

It is therefore clear that a Wheatstone bridge circuit must in practice always include a potentiometer to balance the bridge, compensating for all the various disturbing effects. A typical set up is represented in Figure 3.10.

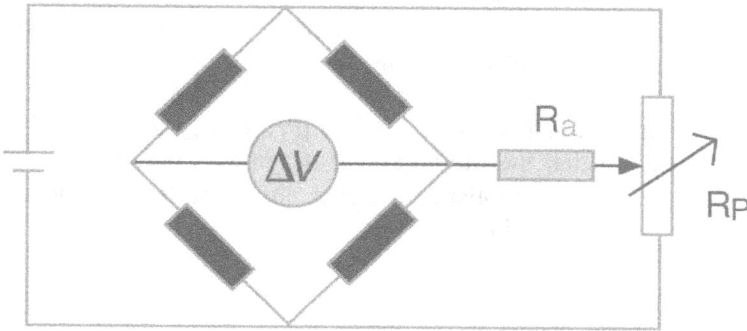

Figure 3.10. Typical set up with potentiometer for fine tuning of the bridge balance.

A potentiometer R_p of 50 kΩ, 10 turn model, and a resistance R_a of 30 kΩ are generally used.

This fine tuning method necessarily implies a decrease in the bridge sensitivity.

An alternative solution without this drawback consists of the addition of a symmetrical voltage at the amplifier stage.

The output signal of a strain gauge bridge has very low amplitude, as we have seen. High resolution measurement equipment is therefore required or, alternatively, an amplification system has to be included in the signal conditioning.

On the other hand it may be convenient that the Wheatstone bridge output signal be adjustable so that the value output by the measurement system can be directly related to the measured strain. A variable gain amplifier can be used for this purpose.

A few numbers !

Consider a quarter bridge circuit whose resistances are equal to the nominal value of the strain gauge, with a voltage supply of 1.5 V.

Admit that the mechanical stress produces a relative variation of 2000×10^{-6} in the strain gauge resistance. If the *GF* is 2.00 this corresponds to a strain ε of 1000 $\mu\varepsilon$.

The bridge unbalance is in this case 750 μV. The system amplification may be tuned so that the output signal becomes, for example, 1 V or 1000×10^{-6} V, which establishes a direct correspondence between the readout and the strain value.

A system with these tuning features should also include the possibility of adjusting the *GF* value supplied by the strain gauge manufacturer.

The amplifiers incorporated in the signal conditioning of these systems shall therefore provide tuning capabilities for various parameters in addition to possessing other features inherent to instrumentation amplifiers.

In conclusion, static strain measurement requires instrumentation with good stability, accuracy and high resolution, with analogue or digital readout and signal output acquisition facilities. Multi-point, manual or automatic, measurement capabilities are also frequently required.

In dynamic measurement operations, namely above 0.1 Hz, the amplifier should have adequate frequency response and automatic channel selection.

In fact commercially available strain gauge measurement bridges have sophisticated electronic conditioning equipment with several tuning capabilities, quarter-, half- and full-bridge selection with high stability power supply and with pre-settings for the most common nominal resistance strain gauge values.

As an example Figure 3.11 shows the control panel of an RDP Electronics M600 amplifier.

Figure 3.11. M600 Amplifier – RDP Electronics.

One can easily identify potentiometers for bridge balancing, for adjusting GF from 1.00 to 10.00 for a direct reading of the output signal in terms of strain (from 100 $\mu\varepsilon$ to 100 k$\mu\varepsilon$), etc.

Chapter 4

Fibre Optic Strain Gauges[1]

4.1. Introduction

Optical fibre sensors may be defined as devices through which a physical, chemical, biological, or other measurand interacts with light, either guided in an optical fibre (intrinsic sensor) or guided to an interaction region (extrinsic sensor) by an optical fibre, to produce an optical signal related to the parameter of interest, Figure 4.1.

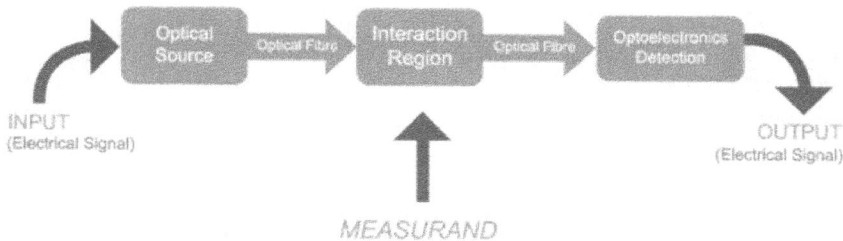

Figure 4.1. Principle of the optical fibre sensor: the parameter under measurement (measurand) changes the properties of the light that propagates in the fibre or exits the fibre in the interaction region.

Fibre sensors can be designed so that the measurand interacts with one or several optical parameters of the guided light (intensity, phase, polarization and wavelength). Independently of the sensor type, the light modulation must be processed into an optical intensity signal at the receiver, which subsequently performs a conversion into an electrical signal. In general, the main interest in this type of sensors comes from the fact that the optical fibre itself offers numerous operational benefits. It is electromagnetically passive, so it can operate in high and variable electric field environments (like those typical of

[1] José Luís Santos, L. A. Ferreira and F. M. Araújo (invited authors).

the electric power industry) and where there is explosion risk; it is chemically and biologically inert since the basic transduction material (silica) is resistant to most chemical and biological agents; its packaging can be physically small and lightweight. Considering the intrinsic low optical attenuation of the fibre (around 0.2 dB/km), it is possible to attain distributed sensing, i.e. determine the measurand as a function of the position along the length of the fibre interrogating from only one end. Also, the optical fibre can be operated over very long transmission lengths, so the sensor can easily be placed kilometers away from the monitoring station. Adding to this, as shown in Figure 4.2, it is also possible to perform multiplexed measurements using large arrays of remote sensors, operated from a single optical source and detection unit, with no active optoelectronic components located in the measurement area, thereby retaining electromagnetic passiveness and environmental resistance.

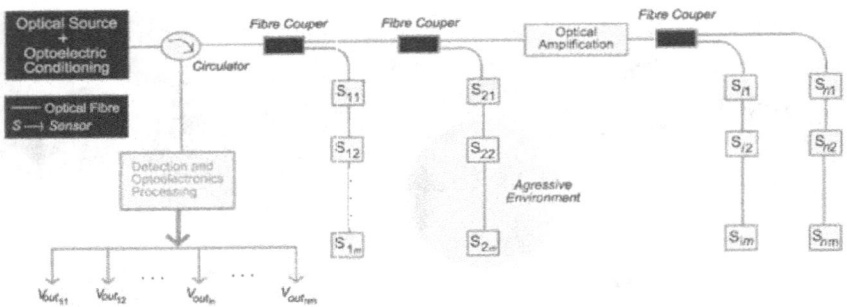

Figure 4.2. A possible architecture for multiplexing of fibre optic sensors.

There are several mechanisms that can be used to measure strain with optical fibres. Here we consider intrinsic sensing, i.e., situations where the measurand interacts with light that keeps propagating in the optical fibre. In such cases, the measurement principle stands invariably in the strain induced variation of the optical path, i.e., the variation of the product nL, where n is the refractive index of the optical fibre core and L is the fibre length between the fixing points (gauge length). Within this sensing concept two approaches will be presented: one involves interferometry and the other is based on the properties of fibre Bragg gratings.

4.2. Interferometric Strain Sensing

There are several optical interferometric configurations that can be used for strain sensing. As an illustrative example, Figure 4.3 shows the Mach-Zehnder configuration.

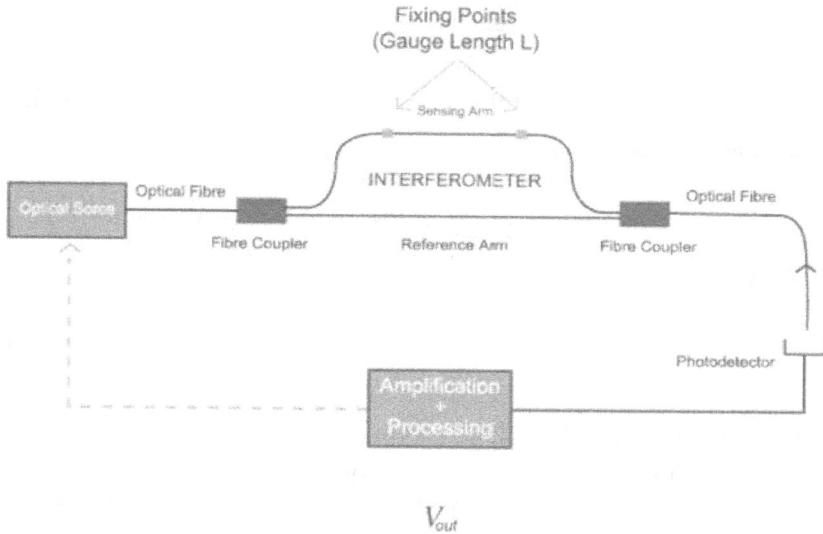

Figure 4.3. Mach-Zehnder interferometric configuration.

The optical source is typically pigtailed to an optical fibre, as well as the photodetector. Therefore, from the optical source to the photodetector one have an all-fibre set-up. The interferometer is formed by two fibre lengths (the sensing and reference arms) and two couplers (with a standard 50/50 splitting ratio). The first one splits the light coming from the optical source into the two arms, while the second one implements the recombination of the light waves that propagated in the arms. If the optical source has adequate coherence properties (such as a laser), these two light waves "recognize" each other and interference occurs. The features of this interference depend on the history of the propagation of the light in each of the arms. Therefore, if one of them (reference arm) is kept unperturbed, then interference has embedded information about the perturbations coupled to the sensing arm. If strain is applied to this arm, the light that reaches the photodetector shows changes that are dependent on the measurand interaction. With adequate signal processing, which may require the implementation of a feedback loop to modulate/tune the emission characteristics of the optical source, it is possible to obtain an output voltage proportional to the applied strain.

The intensity of light that exits the interferometer and reaches the photodetector is given by

$$I_{out} = I_0 \left[1 + V \cos \Phi \right],$$
(4.1)

where I_0 is proportional to the optical power injected into the fibre system, and the fringe visibility, V, and phase, Φ, are given by:

$$V = \frac{I_{max} - I_{min}}{I_{max} + I_{min}}$$
(4.2)

$$\Phi = \frac{2\pi n \Delta \ell}{\lambda}$$
(4.3)

In equation (4.2) I_{max} (I_{min}) is the maximum (minimum) optical intensity that reaches the photodetector. When there is no interference, $V = 0$. Therefore, the visibility identifies the degree of correlation between the two interfering waves and depends on the optical source properties (coherence), as well as on the relative intensities and polarization states of the two interference waves. In equation (4.3), λ is the wavelength of light and $\Delta \ell$ is the geometrical path imbalance of the two interferometer arms. When the product $n \Delta \ell$ changes (for example, due to applied strain), the phase also changes, as well as the output intensity. Due to the small value of the wavelength of light (around 1 μm), small changes in $n \Delta \ell$ translate into noticeable changes in phase. This is the intrinsic reason why optical interferometry turns possible highly sensitive measurements.

Equation (4.1) also shows that the interferometer output depends in a non linear way on the optical phase; moreover, due to the periodicity of the cosine function, there is also ambiguity when one tries to obtain the phase directly from I_{out}. Therefore, the recovery of Φ from I_{out} is not a straightforward process, and complex techniques have been developed along the years aiming at this objective. Globally, this process is identified as interferometric interrogation.

4.2.1. Interferometric Strain Sensitivity

From equation (4.3), the optical phase associated to the propagation in the gauge length L, Figure 4.3, is given by:

$$\phi = \beta L, \quad \beta = \frac{2\pi n}{\lambda} \qquad (4.4)$$

When strain is applied, the phase changes by an amount

$$\Delta\phi = \beta\Delta L + L\Delta\beta, \qquad (4.5)$$

where $\beta\Delta L$ is the phase shift corresponding to the physical change in length of the fibre resulting from axial strain, and $L\Delta\beta$ is relative to the phase change produced by the stress-induced changes in the fibre propagation constant:

$$L\Delta\beta = L\frac{\partial\beta}{\partial n}\Delta n + L\frac{\partial\beta}{\partial a}\Delta a \qquad (4.6)$$

In this equation a is the fibre diameter and Δa is the diameter change due to the applied strain. The term $L(\partial\beta / \partial a)\Delta a$ corresponds to the phase change induced by modal dispersion resulting from a change in fibre diameter. In general it is quite small and can be neglected. On the other hand, $L(\partial\beta / \partial n)\Delta n$ describes the phase change produced by the strain-optic effect (elasto-optic effect: variation of the fibre refractive index due to the applied strain). Therefore, equation (4.5) can be written:

$$\Delta\phi \approx \beta\Delta L + L\frac{\partial\beta}{\partial n}\Delta n \qquad (4.7)$$

When the fibre is subjected to uniform longitudinal strain, ε, then the strain vector is

$$S = \begin{bmatrix} \varepsilon \\ -\nu\varepsilon \\ -\nu\varepsilon \end{bmatrix} \qquad (4.8)$$

where the axial direction of the fibre is the x axis and v is the Poisson ratio. It can be shown that, in these conditions, the strain sensitivity of the fibre is given by

$$\frac{\Delta\phi}{\varepsilon L} = \beta\left[1 - p_\varepsilon\right] \qquad (4.9)$$

$$p_\varepsilon = \frac{n^2}{2}\left[\left(1 - v\right)P_{12} - vP_{11}\right] \qquad (4.10)$$

The parameters P_{11} and P_{12} are the Pockels coefficients (p_ε is the effective photoelastic constant). For a typical fibre, one has:

$$
\begin{aligned}
n &= 1.465 \\
v &= 0.17 \\
P_{11} &= 0.12 \\
P_{12} &= 0.27
\end{aligned}
\Rightarrow \quad \frac{\Delta\phi}{\varepsilon L} \approx \beta\left[1 - 0.22\right] \qquad (4.11)
$$

This result indicates that the fibre length variation due to the applied strain has a much larger effect than the one associated with the strain-induced refractive index variation.

For light with wavelength $\lambda = 1550$ nm,

$$\frac{\Delta\phi}{\varepsilon L} \approx 4.7 \times 10^6 \, rad / m \qquad (4.12)$$

If the gauge length $L = 1$ m and $\varepsilon = 1$ microstrain, then $\Delta\phi = 4.7$ rad/m. This is quite a large phase change and, therefore, it can be asked what would be the minimum strain change that can be detected by the optical system. To answer this question it is necessary to specify the minimum detectable phase shift, which is a function of the demodulation approach and level of noise in the system. For quasi-static variations, as is the case, it is relatively straightforward to detect a phase change of 1 mrad. This means that, for the gauge length considered, $\Delta\varepsilon_{min} = 0.2$ nanostrain, which corresponds to a minimum detectable fibre length variation of 2 Angstroms. This value is of the order of the size of the atom, so it is interesting to think how it is physically possible to measure such small displacements if it is taken into account the random atomic movement due to a none zero temperature. A clue comes from the average effect, which filters out the "fast movement" leaving only the tiny collective average displacement.

These numbers indicate that it is possible to measure very small strain changes with optical interferometry. However, due to the periodic characteristic of the interferometer transfer function, the measurement of absolute strain values is not straightforward. This means that if the sensing system goes down, when it is turned on again the absolute strain value can be lost, being only possible to detect strain changes after this moment. There are optical processing techniques that can circumvent this limitation, however at the price of an increased complexity. It is also important to mention the cross-sensitivity effect of temperature, which is very strong in optical fibre sensing interferometry. This means that the utilization of this concept to measure strain is normally only considered in very demanding situations, where the critical issue is to detect the minimum possible strain changes without too many constraints about its implementation cost (as is the case in the large optical interferometry experiments that are under way to measure the very small strain changes that are expected to be produced in the interferometer arms due to cosmic gravitational waves).

Nowadays, strain measurement with optical fibres is almost exclusively done using a notable fibre optic component: the Fibre Bragg Grating.

4.3. Strain Sensing with Fibre Bragg Gratings

4.3.1. Introduction to Fibre Bragg Gratings

Fibre Bragg gratings (FBG) are simple, versatile and small intrinsic sensing elements that can be written in silica fibres, and which consequently have all the advantages normally attributed to fibre sensors. In addition, due to the fact that the measurand information is encoded in the resonant wavelength of the structure, which is an absolute parameter, these devices are inherently self-referenced and can be easily multiplexed, which is particularly important in the context of distributed sensing. All these characteristics triggered a research burst by the mid-nineties, addressing diversified topics like the fundamentals of UV induced refractive index modulation of the fibre core, interrogation of these wavelength encoded devices, and new sensing head concepts integrating FBG, including their multiplexing, and applications.

Figure 4.4 a) shows the structure and spectral characteristics of these devices. The fibre gratings are fabricated UV imprinting an axial periodic pattern in the fibre optic core of higher and lower refractive index modulation (in the sections of UV irradiation, the fibre glass is modified and, under certain circumstances, the refractive index also changes - a process called photosensitivity). When broadband light (with a large spectrum) propagates down the fibre and reaches the FBG position, at each refractive index interface there will be a tiny reflection (Fresnel reflection), generating very small amplitude reflected waves (the amplitude is small because the amplitude of the refractive index modulation is also small – typically around 1×10^{-3}). For the large majority of the incident wavelengths (such as λ_1 in Figure 4.4b)) these reflected waves are out of phase and, therefore, when they add up the result is destructive interference and no light is reflected, i.e., all light is transmitted (it is assumed that the number of small reflected waves is large; typically, the refractive index modulation period is around 0.5 µm, so for a grating length of 10 mm, this means that we have ~20 000 reflected waves, which fulfils the required condition). However, there is a small wavelength window where the reflected waves interfere in phase (around the wavelength λ_2 in Figure 4 b). In these cases the interference is constructive and a strong reflection occurs. The central wavelength of this spectral window satisfies the so called Bragg condition

$$\lambda_B = 2n\Lambda \,, \tag{4.13}$$

where n is the refractive index of the fibre core and Λ is the period of the refractive index modulation, Figure 4.4 a). The reflected wavelengths are in a spectral window with a width of approximately 0.1 nm to 0.2 nm around λ_B All other wavelengths are transmitted, i.e., they behave as if the refractive index modulation of the fibre core is not present.

Equation (4.13) indicates the crucial feature of these structures: changes in the period Λ, or in the refractive index of the core, n, originate a shift in λ_B, i.e., a small variation occurs in the "colour" of the reflected light. The measurement of this variation gives indications on the action that introduced changes in (n, Λ). Because the wavelength ("colour") is an absolute parameter, this process is insensitive to variations that may occur in other light parameters along the optical system (such as intensity, phase and polarization). This feature and the possibility of having a large number of FBG structures along the same fibre, each one with its own λ_B, brought a qualitative advance to the optical fibre sensing domain.

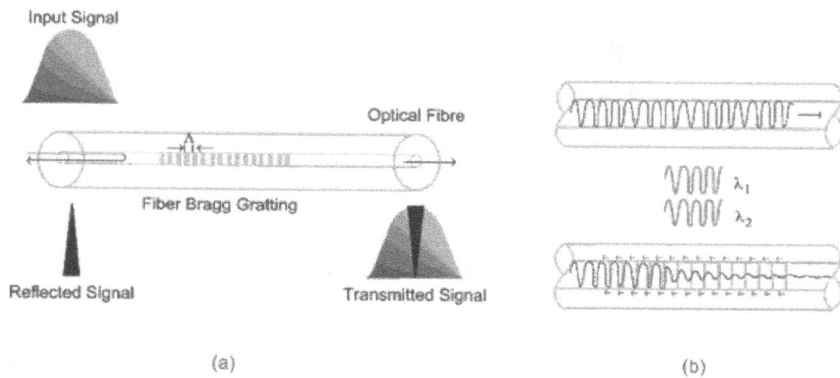

Figure 4.4. Fibre Bragg grating spectral signature (a), and physical principle (b).

As a note, it can be mentioned that FBG interrogation is the designation commonly used to refer ways to convert the Bragg wavelength value (and variations) into an electrical signal with adequate characteristics to obtain the information about the measurand. The general principle of FBG interrogation is shown in Figure 4.5.

The optical source can be a broadband source (LED, SLD, ASE, Supercontinuum), in which case it operates passively and normally there is no control from the processing unit. This is not the case when spectrally narrow illumination is used, most of the cases from laser sources, in which the wavelength modulation of the emitted light can be a component of the FBG interrogation technique.

4.3.2. Strain Sensing

Strain sensing with FBG is based on the variation of λ_B due to this parameter.

73

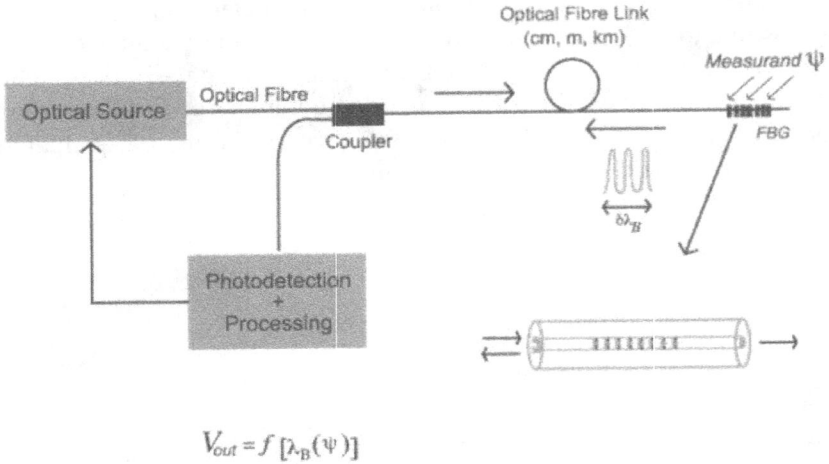

$$V_{out} = f\left[\lambda_B(\psi)\right]$$

Figure 4.5. General layout for interrogation of fiber Bragg grating sensors.

The characterization of this measurement process follows a rational close to the one presented above for the Mach-Zehnder interferometer. The starting point is equation (4.13), from where

$$\frac{\delta\lambda_B}{\lambda_B} = \left(\frac{1}{\Lambda}\frac{\partial\Lambda}{\partial\varepsilon} + \frac{1}{n}\frac{\partial n}{\partial\varepsilon}\right) \qquad (4.14)$$

The first term corresponds to the change in the grating spacing, while the second is related to the elasto-optic effect. Applying the arguments presented before it comes out:

$$\frac{\delta\lambda_B}{\lambda_B} = \left(1 - p_\varepsilon\right)\delta\varepsilon \qquad (4.15)$$

with $p_\varepsilon \approx 0.22$ given by (10). For operation at $\lambda_B = 1550$ nm, it turns out

$$\frac{\delta\lambda_B}{\delta\varepsilon} \approx 1.2\,pm\,/\,\mu\varepsilon \qquad (4.16)$$

If a specific FBG interrogation system is able to read 1 pm, this means the system has a strain resolution of ≈ 1 µε.

4.3.3. Cross-sensitivity to Temperature

As all sensing technologies, fibre optic sensing deals also with the problem of cross-sensitivity. When the targeted measurand is strain, temperature is the critical cross-sensitivity parameter. Focusing on sensing with fibre Bragg gratings, this effect is substantial because the temperature sensitivity of these devices is large (typically ≈ 10 pm/°C). Therefore, it is necessary to perform actions to deal with this problem.

In the large majority of applications, what is done when measuring strain with an FBG is to add a second strain-free FBG, therefore only sensitive to temperature. Its reading allows compensating quite effectively for the temperature effects. However, there are other conceptually more attractive solutions that fit into the broad topic of simultaneous measurement of several parameters.

The principle behind the sensing head solutions that allow simultaneous measurement of a number N of parameters is the identification of N characteristics of the sensing head structure that change differently under the action of the measurands of the set. If this happens, and in the particular but important case of linear dependences, it is always possible to write N independent equations that generate explicit solutions for the actual value of each measurand, even in the situation where all of them are changing. This concept is better illustrated when the simplest case of only two measurands is considered, as it happens for the case under concern here, i.e., strain-temperature discrimination.

In order to write the measurand interaction equations, it will be assumed that the sensing head structure is based on FBG, obviously not a necessary condition. In such case, under variations of applied strain, $\Delta\varepsilon$, and temperature, ΔT, the associated variations of the two required wavelength resonant signatures ($\Delta\lambda_{Bi}$, i = 1, 2) can be written as

$$\Delta\lambda_{Bi} = K_{Ti}\Delta T + K_{\varepsilon i}\Delta\varepsilon \qquad (4.17)$$

The thermal sensitivity, K_{Ti}, depends both on the thermal expansion of the fibre and on the thermo-optic coefficient of the fibre material. As indicated in the previous point, the strain sensitivity, $K_{\varepsilon i}$, depends on the photoelastic coefficient of the fibre, but is mainly determined by the variation of the grating pitch when strain is applied, which is determined by the mechanical properties of the fibre.

For the ideal case of simultaneous measurement of strain and temperature, the wavelengths depend only on one of the parameters, i.e., $\lambda_1 = \lambda_{B1}(T)$ and $\lambda_2 = \lambda_{B2}(\varepsilon)$ – it could be the other way around. In such case, equation (4.17) reduces to the straightforward form

$$
\begin{bmatrix} \Delta T \\ \Delta \varepsilon \end{bmatrix} = \frac{1}{K_{\varepsilon 2} K_{T1}} \begin{bmatrix} K_{\varepsilon 2} & 0 \\ 0 & K_{T1} \end{bmatrix} \begin{bmatrix} \Delta \lambda_{B1} \\ \Delta \lambda_{B2} \end{bmatrix}
$$

(4.18)

The non ideal case is when both wavelengths depend on the two measurands, and equations (4.17) must be processed in order to have an explicit matrix equation that permits to obtain simultaneously ΔT and $\Delta \varepsilon$:

$$
\begin{bmatrix} \Delta T \\ \Delta \varepsilon \end{bmatrix} = \frac{1}{D} \begin{bmatrix} K_{\varepsilon 2} & -K_{\varepsilon 1} \\ -K_{T2} & K_{T1} \end{bmatrix} \begin{bmatrix} \Delta \lambda_{B1} \\ \Delta \lambda_{B2} \end{bmatrix}
$$

(4.19)

$$
D = K_{T1} K_{\varepsilon 2} - K_{\varepsilon 1} K_{T2}
$$

This result turns clear the requirement for this process to work properly, i.e., $D \neq 0$. Out of the ideal case, the most favourable situation occurs when one of the selected characteristics of the sensing head is not affected by one of the measurands.

This concept of simultaneous measurement of strain and temperature has been demonstrated for several fibre optic based sensing heads. The results obtained so far indicate the viability of simultaneous determination of these parameters with resolutions of few $\mu\varepsilon$ and some tenths of degree centigrade, for strain and temperature, respectively.

4.4. Impact of Strain Sensing with Fibre Bragg Gratings

In the context of strain measurement, the fibre Bragg grating can be though to be the optical equivalent of the electrical strain gauge. The application of this new technology in several engineering domains is ongoing at an increasingly fast rate. Information about details of this process can be found in literature, and also in the web pages of the companies fabricating FBG based sensors and related instrumentation. Table 4.1 lists some of these companies.

Table 4.1. Companies fabricating fibre Bragg sensors and instrumentation.

Company	URL address
FiberSensing (Portugal)	http://www.fibersensing.com
Micron Optics (USA)	http://www.micronoptics.com
Insensys (UK)	http://www.insensys.com
FOS&S (Belgium)	http://www.fos-s.com
Ibsen Photonics (Denmark)	http://www.ibsen.dk

As an example, Figure 4.6 shows a strain sensor based in a fiber Bragg grating with a packaging similar to the one of a standard resistive strain gauge.

Figure 4.6. Strain gauge based in a fiber Bragg grating (source: FiberSensing).

Chapter 5

Uncertainty Evaluation in Stress Measurement[2]

Measurement is fundamental in all engineering areas. But the result of a measurement is only complete when it is complemented with information about its associated uncertainty. An example will be provided in this module using the determination of the normal bending stress on the surface of an aluminium cantilever beam instrumented with resistance strain gauges.

5.1. Basic Concepts

Units of measurement were among the earliest tools invented by human civilization, thousands of years ago. But the result of a measurement is never exact and some degree of uncertainty is always associated to it. So a measurement result is only complete when there is also information about its associated uncertainty. The uncertainty of the result is an essential parameter in order to assess if the result is satisfactory for the intended objective or for evaluating its consistency with other similar results, either in R&D activities or for industrial manufacturing. The uncertainty of a result indicates the level of confidence of its value within the uncertainty interval.

In fact any measurement is always subject to random perturbations, such as temperature and humidity fluctuations or operational variability. Other disturbances may be due to systematic effects such as those associated to the measurement instruments as is the case of an offset or a temperature drift.

The uncertainty of a measurement result is generically made up of several components categorized according to the method used to

---

[2] This chapter is written by Maria Teresa Restivo with collaboration of invited author Carlos Sousa.

evaluate them: type A and type B. Those of the first type are evaluated statistically and are therefore quantified by evaluating repeated measurements. The second type is quantified by means other than the statistical analysis of series of observations. An estimation based on equipment characteristics and results of its calibration, among other items, is done.

Repeated measurement values commonly follow a normal distribution. The statistical parameter relevant for their evaluation is the standard deviation.

The second category of uncertainty components is also expressed by standard deviation, even if not of normal distribution type. For example, if the value of a quantity lies with an equal probability within a certain interval, the standard deviation is estimated by an expression associated to the rectangular distribution type.

Using the root sum square method all the components are associated for producing the combined uncertainty. Usually an expanded standard uncertainty is determined by multiplying the standard uncertainty by a coverage factor. Obviously the greater this factor, the larger is the uncertainty interval and so the higher is the confidence level for the value lying within that interval. Normally, for a coverage factor of 2 the level of confidence is 95 %.

So the uncertainty of a result is an essential attribute and its relevance is even more fundamental when the result is close to one of the limits of a specified interval. Finally, in general terms, its knowledge improves the reliability and quality of a result.

During the 1980s various international workgroups contributed actively for the treatment of measurement uncertainty which led to a general methodology compatible with many different measurement areas. The best known group, named TAG4, presented its work in December 1992, in Barcelona, Spain. This work became the basis of the Guide of Uncertainty of Measurement – GUM. The revised version of 1995 has been used as a reference worldwide. In September 2008 a new revision with minor corrections has been published: "JCGM 100:2008 Evaluation of measurement data — Guide to the expression of uncertainty in measurement".

In the next section an analytical formulation will be presented according this document.

5.2. Evaluation of Measurement Uncertainty: the Modelling Approach

There are different methods for evaluating measurement uncertainties: the modelling approach (a step-by-step approach), the single-laboratory data validation approach and the inter-laboratory data validation approach.

In the example presented below the modelling approach is used.

The modelling approach uses "a model formulated to account for the interrelation of all the influence quantities that significantly affect the measurand. Corrections are assumed to be included in the model to account for all recognised, significant systematic effects. The application of the law of propagation of uncertainty enables evaluation of the combined uncertainty on the result. The approach depends on partial derivatives for each influence quantity", (Eurolab, 2007). This is the more general method. If the modelling is difficult to achieve, then any of the other alternatives should be used.

5.3. Uncertainty Evaluation Using the Modelling Approach

The complete result of a measurement is given by the equation 5.1 considering also Figure 5.1.

$$y' = y \pm U, \qquad (5.1)$$

where y is the measurement result, y' is its complete result and U the expanded uncertainty.

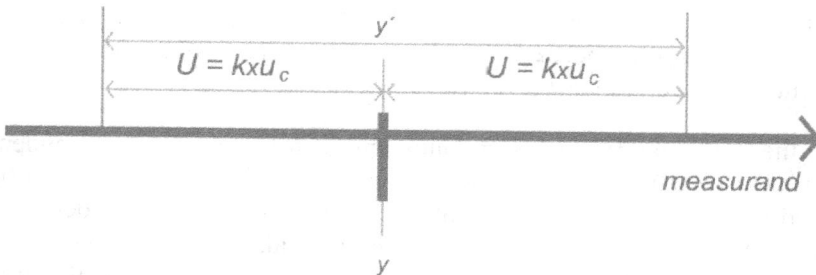

Figure 5.1. Complete expression of a measurement result.

Figure 5.1 permits "to feel" that a measurement always includes in it a central tendency (the average) and a measure of variability or dispersion (standard deviation) and that these concepts do not have a precise position. Since there could be approximations and estimates those could contribute for making "worse" the uncertainty value.

The evaluation of the confidence interval is always enlarged by the coverage factor, k, usually with a value of 2 or even larger. This coverage factor, statistically, brings more confidence to the uncertainty value, which commonly is around 0.95 (95 %). The expanded uncertainty U is given by multiplying k by the combined standard uncertainty, u_c.

$$U = k \times u_c \qquad (5.2)$$

To obtain this final result some essential steps have to be performed.

Usually, the measurement of a quantity is of indirect type as are generally those coming from testing procedures. The measurement of a quantity Y is a function of several input quantities, X_i, and this function provides the model for evaluating the uncertainty associated with Y:

$$Y = f(X_1, X_2, X_3, ... X_n) \qquad (5.3)$$

But, as the quantity values X_i are only estimated and their true values are never known, lower case letters are conventionally used for both the input variables and the output, the measurand:

$$y = f(x_1, x_2, x_3, ... x_n) \qquad (5.4)$$

Once the model is known as well as the estimated values for the input quantities, x_i, the next step consists on the evaluation of the associated uncertainty components, $u(x_i)$ or u_i, for calculating the standard uncertainty $u(y)$ associated with the measurement result, y, also named combined uncertainty, u_c.

If the measurand follows a Gaussian distribution for independent repeated measurements, a coverage factor $k = 2$ is recommended to be used and the expanded uncertainty guarantees a level of confidence of 95 % for the value lying within that interval. These conditions are observed in most calibration procedures. In other cases the effective degrees of freedom should be determined using the t-Student distribution or the *Welch-Satterthwaite*[3] formula.

5.3.1. How to Estimate the Standard Uncertainty of Type A

Whenever it is possible to perform in the same conditions n independent measurements of one of the input quantities x_i, the standard uncertainty of type A associated with the estimated input xi, is given by the experimental standard deviation of the mean, s, divided by the square root of the number of measurements, n:

$$u(x_i) = \frac{s}{\sqrt{n}} = \sqrt{\frac{\sum\limits_{i=1}^{n}(x_i - \bar{x})^2}{n(n-1)}} , \qquad (5.5)$$

where the number of independent measurements should be larger than ten ($n > 10$).

5.3.2. How to Estimate the Standard Uncertainty of Type B

When a statistic analysis is not possible for estimating $u(x_i)$ or u_i of an input quantity x_i, then the standard uncertainty to be evaluated is of Type B. This happens when the quantity values result from previous measurements or are related with manufacturers' specifications, calibration information, hysteresis, etc.

When available information is not enough to describe the distribution type, it is prudent to consider a uniform distribution within an interval, called of Type B. The more common ones are presented at Table 5.1.

---

[3] EA-4/02 - Expression of the Uncertainty of Measurement in Calibration,
EA - European Co-operation for Accreditation, December 1999.

Table 5.1. Examples of uncertainty sources of Type B.

Uncertainty source	Distribution type	Standard evaluation
Equipment resolution (r)	B rectangular	$u = r /\sqrt{12}$
Hysteresis (h)	B rectangular	$u = h /\sqrt{12}$
Maximum admissible error (MAE): $\pm a$	B rectangular	$u = a /\sqrt{3}$

5.3.3. Law of Uncertainty Propagation

The standard uncertainty of the result, $u(y)$, is the positive square root of the estimated variance, $u^2(y)$, given by equation 5.6:

$$u^2 (y) = \sum_{i=1}^{N} \left(\frac{\partial f}{\partial x_i} \right)^2 u^2 (x_i) + 2 \sum_{i=1}^{N-1} \sum_{j=i+1}^{N} \frac{\partial f}{\partial x_i} \frac{\partial f}{\partial x_j} u(x_i, x_j) \quad (5.6)$$

Equation 5.6 is derived from a first-order Taylor series approximation of equation 3, and is known as the law of propagation of uncertainty. If there is no correlation between the input quantities, then the standard uncertainty of the result, u(y), is:

$$u(y) = \sqrt{\sum_{i=1}^{N} \left(\frac{\partial f}{\partial x_i} \right)^2 u^2 (x_i)} = \sqrt{\sum_{i=1}^{N} u^2 (x_i) \cdot c_i^2}, \quad (5.7),$$

where c_i are the partial derivatives $\partial f / \partial x_i$ also named sensitivity coefficients, since those derivatives convey the influence of each standard uncertainty $u(x_i)$, associated with the input estimate x_i, on the final measurement standard uncertainty, $u(y)$, or u_c.

The expanded uncertainty, according to equation 5.2, is then obtained multiplying the final measurement standard uncertainty of the result, $u(y)$, by the coverage factor, k.

5.4. Evaluation of Uncertainty Associated to Stress Measurement

For this study an aluminium cantilever beam of constant rectangular cross section has been tested in bending. The test specimen was instrumented with two electrical resistance strain gauges, symmetrically bonded respectively on the top and bottom surfaces and oriented along the longitudinal axis of the beam. The strain gauges are integrated in a half Wheatstone bridge circuit. The load is applied close to the free end of the beam, at a distance L from the geometric centre of the strain gauges grid, Figure 5.2.

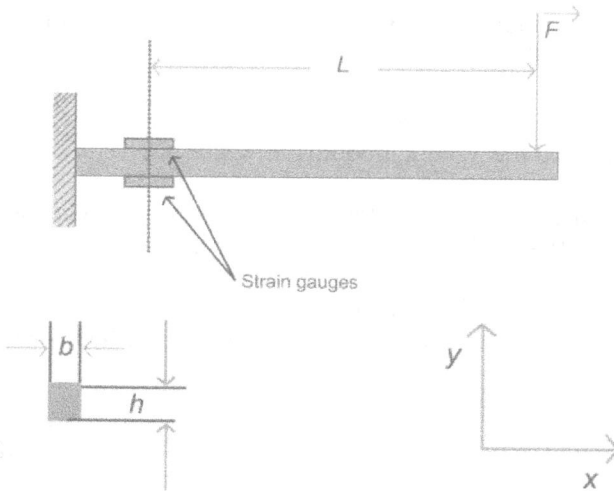

Figure 5.2. Cantilever beam.

In a uniaxial tensile test within a moderate stress range many materials exhibit a linear relation between the axial stress and strain which can be expressed by Hooke's law:

$$\sigma = E\,\varepsilon, \qquad\qquad (5.8)$$

where E is the Young modulus of the material.

The measuring system used in this study is represented by the block diagram of Figure 5.3:

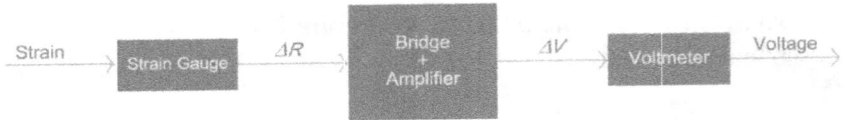

Figure 5.3. Block diagram of the strain measuring system.

The stress in the bonded strain gauge grid area does not affect directly the resistance strain gauges. They are sensitive to the strain which can be measured through the bridge output voltage as a result of changes in the strain gauge resistance. So it is important to know the selected bridge scale and its calibration. In the present case: 1 mV/μm/m.

The Young modulus determined in a previous test is (65.88 ± 0.84) GPa.

Using equation 5.8, it is possible to calculate the stress, σ_x, once known the strain, ε_x, and the Young modulus, E, $(x_i: \varepsilon, E)$.

The uncertainty associated to the strain measurement depends only on the characteristics of the instrumentation used.

For a better perception on the method used for estimating the uncertainty a *"cause-effect"* diagram has been sketched for the present example, Figure 5.4.

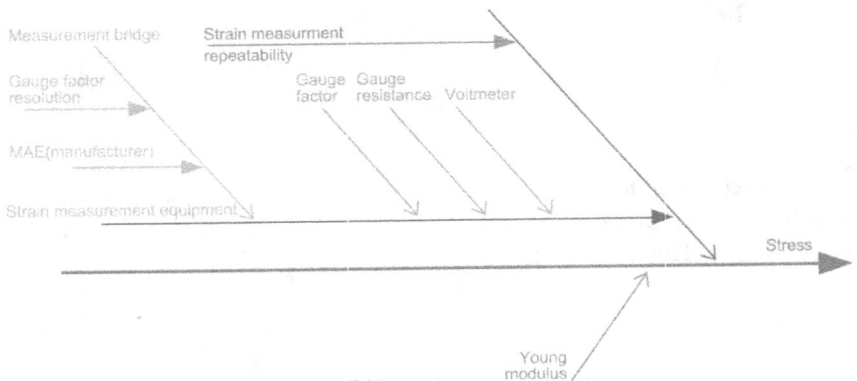

Figure 5.4. *"Cause-effect"* diagram for the model used in the evaluation of the stress measurement uncertainty.

For the evaluation of the stress uncertainty it is also very important to consider the contribution from equipment (bridge and voltmeter) and sensors, either using calibration certificates or, at least, manufacturing characteristics. As an example, Figure 5.5 presents the strain gauge data sheet.

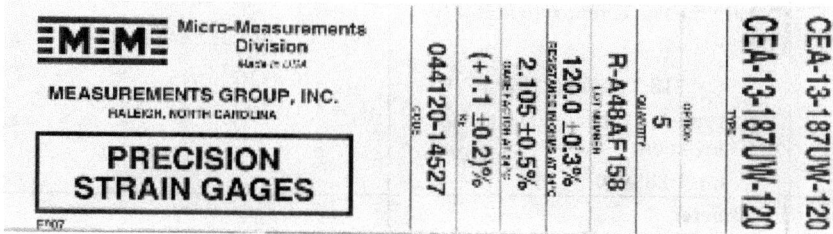

Figure 5.5. Strain gauge data sheet.

Table 5.2 summarizes the components of the input data, either those related with the equipment and the measurement repeatability or with Young modulus value and its associated uncertainty.

Accounting for these components also involves determining the respective sensitivity coefficients related to each input variable (x_i: E and ε). From equation 5.8 we get:

$$\partial\sigma/\partial E = \varepsilon, \quad \text{m/m} \tag{5.9}$$

$$\partial\sigma/\partial\varepsilon = E, \quad \text{N m}^{-2} \tag{5.10}$$

Considering that the gauge factor, *GF*, is given by

$$GF = (dR/R)/(dL/L) = (dR/R)/\varepsilon \tag{5.11}$$

the strain, ε: is also related with *GF* and the strain gauge resistance, *R*.

One should therefore consider the contributions of the rectangular uncertainties Type B related with the tolerances inherent to the sensors (see sensor manufacturer data sheet in Figure 5.5) and to the gauge factor tuning resolution. The sensitivity coefficients related to the variables *R* and *GF* are given as:

$$\partial\varepsilon/\partial GF = -\varepsilon/GF \quad \text{(dimensionless quantity)} \tag{5.12}$$

$$\partial \varepsilon / \partial R = -\varepsilon / R, \quad \Omega^{-1} \qquad (\,5.13\,)$$

Table 5.2. Uncertainty components for evaluating the uncertainty of the stress.

i	Physical quantity / uncertainty component	Standard uncertainty
1.	**Strain measurement repeatability (mean)** $\overline{\varepsilon} = 218.2$ µm/m (2182×10^{-7}) Type A, experimental standard deviation $s_\varepsilon = 1.888 \times 10^{-7}$	$u_\varepsilon = 1.888 \times 10^{-7} / \sqrt{13}$ $u_\varepsilon = 5.24 \times 10^{-8}$
2.	**Voltmeter** Measured value = 0.2182 V Range = 1 V Maximum admissible error (manufacturer): ± 30 ppm reading ± 5 ppm range Type B, rectangular distribution.	$u_V = (30 \times 10^{-6} \times 0.2182 + 5 \times 10^{-6} \times 1)/\sqrt{3}$ $u_V = 6.67 \times 10^{-6}$ V
3.	**Measurement Bridge** Range= 40 µm/m Maximum admissible error (manufacturer): ±0.02 % range (range: 10^4 µm/m) Type B, rectangular distribution.	$u_P = (0.02/100 \times 10^4 \times 10^{-6})/\sqrt{3}$ $u_P = 1.16 \times 10^{-6}$
4.	**Strain Gauge Resistance R** Maximum admissible error (manufacturer): ± 0.3 % of nominal value (120 Ω). Type B, rectangular distribution.	$u_R = (120 \times 0.3/100)/\sqrt{3}$ Ω $u_R = 2.08 \times 10^{-1}$ Ω
5.	**Gauge Factor GF** Maximum admissible error (manufacturer): ± 0.5 % nominal value (GF = 2.1). Type B, rectangular distribution.	$u_{GF} = (2.1 \times 0.5/100)/\sqrt{3}$ $u_{GF} = 6.06 \times 10^{-3}$
6.	**Young modulus** Test result value = 65.88 GPa Expanded uncertainty = ± 0.84 GPa Coverage factor k = 2.00 Type B, Gaussian distribution	$u_E = 0.84 \times 10^9 / 2$ Pa $u_E = 0.42 \times 10^9$ Pa

Tables 5.3 and 5.4 present the uncertainty components associated to the equipment and sensors (measuring bridge, voltmeter, strain gauge resistance and gauge factor) as well as the uncertainty associated to the stress and to the estimation of its expanded uncertainty.

Table 5.3. Uncertainties of equipment and sensors.

	Component	evaluation	unit	type	$u(x_i)$	c_i	$u^2_i(y)$	GL
	Equipment uncertainty							
1	GF	2,105	dimensionless	BR	6,077E-03	-1,037E-04	3,968E-13	500
2	Resistance	120	ohm	BR	0,2078	-1,81BE-05	1,428E-13	500
3	Voltmeter	2,18E-01	volt	BR	6,666E-06	1,00E-03	4,444E-17	500
4	Bridge		dimensionless	BR	1,155E-06	1	1,333E-12	500
5	Bridge resolution		dimensionless	BR	0,005773503	-1,037E-04	3,582E-13	500
						$u^2(y)$	2,231E-12	
						$u(y)$	1,494E-06	
						k	2	
						$U=$	3,0E-06	

Table 5.4. Stress uncertainty and evaluation of expanded uncertainty.

i	Component	Evaluation	Unit	Type	$u(x_i)$	c_i	$u^2_i(y)$	GL
	Uncertainty							
1	Strain (equipment)	2,18-04	dimensionless	DN	1,49E-06	6,59E+10	9,60E+09	50
2	Strain (repeatability)	6,59C+10	dimensionless	A	5,05E-08	6,59E+10	1,11E+07	13
3	Young modulus		Pa	DN	4,20E-08	2,18E-04	0,40E+09	50
	Stress -	1,438E+07	Pa			$u^2(y)$	0,41E+09	Pa^2
	Stress-	14,38	MPa			$u(y)$	91704	
						GL(eff)	21,52	Pa
						k	2,13	
						$U=$	1,9E+05	Pa
						$U=$	0,19	MPa
						$U(rel)=$	1,4	%

A more intuitive way of representing the results from Tables 5.3 and 5.4 is illustrated in Figure 5.6.

The final result obtained for the stress and the corresponding expanded uncertainty is:

$$\sigma = (14.38 \pm 0.19)\ MPa$$

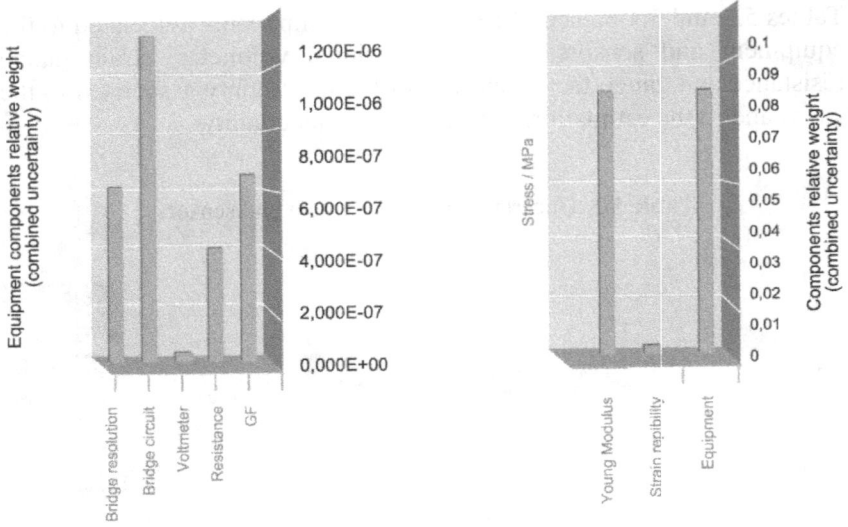

Figure 5.6. Graphical representation of the relative weight of the associated uncertainty components.

5.5. Final Remarks

The weight of the uncertainty component associated to the measuring bridge is the most significant, although very close to that for the Young modulus. In the example presented the analytical model was very simple for the sake of clarity. However the method can be applied to much more complex situations.

It is interesting to note that in spite of the large contribution from the equipment for the final stress uncertainty evaluation, the strain measurement repeatability value is good. So if the criterion was used of considering the measure of variability or dispersion as interval of confidence, a better (lower) uncertainty would be incorrectly found.

Using the information for the uncertainty component for strain repeatability (i = 2 in Table 5.4), $u_i^2(y) = 1.11 \times 107$ Pa^2, we get an expanded uncertainty:

$$U = \pm((1.11 \times 10^7)^{\frac{1}{2}} \times 2) \text{ Pa} = 6.7 \times 10^3 \text{ Pa} = 0.0067 \text{ MPa}.$$

This simply means that the expanded uncertainty would be around 30 times lower than that evaluated using the modelling approach (0.19 MPa).

Bibliography

A. Bray, G. Barbato, R. Levi, Theory and Practice of Force Measurement, *Academic Press*, ISBN 0-12-128453-0, 1990.

A. L. Window, G. S. Holister, Strain Gauge Technology, *Kluwer Academic Publishers Group,* ISBN: 9781851668649, 1992.

Akhtar S. Khan and Xinwei Wang, Strain Measurements and Stress Analysis, *Prentice-Hall Inc.,* ISBN 0-13-080076-7, 2001.

Alan S. Morris, Measurement and Instrumentation Principles, *Elsevier,* Butterworth- Heinemann, ISBN 0 7506 5081 8, 2001.

Aurélio Campilho, Instrumentação Electrónica. Métodos e Técnicas de Medição, *FEUP Edições,* ISBN 972-752-042-1, 2000.

Cusano, A. Cutolo, J. Albert (Editors), Fiber Bragg Grating Sensors: Recent Advancements, Industrial Applications and Market Exploitation, *Bentham*, 2011.

D. A. Krohn, Fiber Optic Sensors. Fundamentals and Applications, *Instrument society of America*, ISBN 0-55617-010-6, 1991.

EA Guideline EA-4/16 Expression of Uncertainty in Quantitative Testing EA2003.

Engineering Measurements, *John Wiley and Sons Inc.,* ISBN 0-471-55192-9, 1993.

Epsilonics, Photoelastic Methods – Reflections or Transmission?, *The Measurement Group Journal for Stress Analysts,* Vol. II, Issue 3, USA, December 1982.

Epsilonics, *The Measurement Group Journal for Stress Analysts*, Vol. II, Issue 2, USA, July, 1982.

Ernest O. Doebelin, Measurement Systems, Applications and Design, *MacGraw-Hill*, ISBN 0-07-100697-4, 1990.

Expression of the Uncertainty of Measurement in Calibration EA-4/02, December 1999.

F. Ansari, Y. Libo, Mechanics of Bond and Interface Shear Transfer in Optical Fiber Sensors, *Journal of Engineering Mechanics*, 124, 1998, pp. 385-394.

G. B. Hocker, Fiber-Optic Sensing of Pressure and Temperature, *Applied Optics*, 18, 1979, pp. 1445-1450.

G. Gagliardi, M. Salza, S. Avino, P. Ferraro, P. De Natale, Probing the Ultimate Limit of Fiber-Optic Strain Sensing, *Science,* 19, 2010, pp. 1081-1084.

General requirements for the competence of testing and calibration laboratories, EN ISO/IEC 17025, 1999.

Georges Asch et collaborateurs, Les Capteurs en Instrumentation Industrielle, *Dunod*, ISBN 2-10-000220-1, 1991.

Guide to the Expression of Uncertainty in Measurement, 1st corr. Edition, *ISO*, Geneva, 1995.

J. López-Higuera M (Editor), Handbook of Optical Fiber Sensing Technology, *John and Sons Wiley*, 2002.

Jacob Fraden, Handbook of Modern Sensors: Physics, Designs, and Applications, *Springer*, ISBN 0-387-00750-4, 2003.

James W. Dally, William F. Riley, Kenneth G. McConnell, Instrumentation for Engineering Measurements, *John Wiley and Sons Inc.*, ISBN 0-471-55192-9, 1993.

JCGM 100:2008, Evaluation of measurement data — Guide to the expression of uncertainty in measurement, September 2008

Jean Avril, Encyclopedie D'Analyse des Contraintes, Micromeasures, Alpha-Imprime, Dêpot Legal N° 7570, Paris, France, 1984.

K. T. Wan, C. K. Y. Leung, Fiber Optic Sensor for the Monitoring of Mixed Mode Cracks in Structures, *Sensors and Actuators A*, 135, 2007, pp. 370-380.

M. Murray, William R. Miller, The Bonded Electrical Resistance Strain Gage, An Introduction, *Oxford University Press*, ISBN 9780195072099, 1992.

Maria Teresa Restivo and Carlos Sousa, Measurement uncertainties in the experimental field, *Sensors & Transducers*, Vol. 95, Issue 8, August 2008, pp. 1-12.

Measurement uncertainty revisited: Alternative approaches to uncertainty evaluation, *Eurolab*, Technical Report N° 1/2007, March 2007.

O. Frazão, L. A. Ferreira, F. M. Araújo, J. L. Santos, Applications of Fiber Optic Grating Technology to Multi-Parameter Measurement, *Fiber and Integrated Optics*, 24, 227-244, 2005.

Peter Elgar, Sensors for Measurement and Control, *Addison Wesley Longman Limited*, 1998, ISBN 0-582-35700-4.

S. Magne, S. Rougeault, M. Vilela, P. Ferdinand, State-of-Strain Evaluation with Fiber Bragg Grating Rosettes: Application to Discrimination Between Strain and Temperature Effects in Fiber Sensors, *Applied Optics*, 36, 1997, pp. 9437-9447.

S. Pevec, D. Donlagic, All-fiber Long-Active-Length Fabry-Perot Strain Sensor, *Optics Express*, 19, 2011, pp. 15641-15651.

Stephen H. Crandall, Norman C. Dahl e Thomas J. Lardner, An Introduction to the Mechanics of Solids, *McGraw-Hill Inc.*, Tokyo, 1978, ISBN 0-07-066230-4.

S. P. Timoshenko, Strength of Materials, ISBN 0-88275-420-3, *Krieger Publishing Company*, 1976.

Stephen P. Timoshenko e James E. Gere, Mecânica dos Sólidos, Vol. 1, *Livros Técnicos e Científicos Editora S. A.*, Rio de Janeiro, 1983, ISBN 85-216-0247-2.

V. Bhatia, D. Campbell, R. O. Claus, A. M. Vengsarkar, Simultaneous Strain and Temperature Measurement with Long-Period Gratings, *Optics Letters*, 22, 1997, pp. 648-650.

Walter D. Pilkey, Formulas for Stress, Strain, and Structural Matrices, *John Wiley & Sons*, ISBN 0-471-03221-2, 2005.

X. F. Yang, S. J. Luo, Z. H. Chen, J. H. Ng, C. Lu, Fiber Bragg Grating Strain Sensor Based on Fiber Laser, *Optics Communications*, 271, 2007, pp. 203-206.

X. M. Tao, L. Q. Tang, W. C. Du, C. L. Choy, Internal Strain Measurement by Fiber Bragg Grating Sensors in Textile Composites, *Composites Science and Technology*, 60, 2000, pp. 657-669.

Media Files List and Locations

Chapter 1:

Angular_distortion:

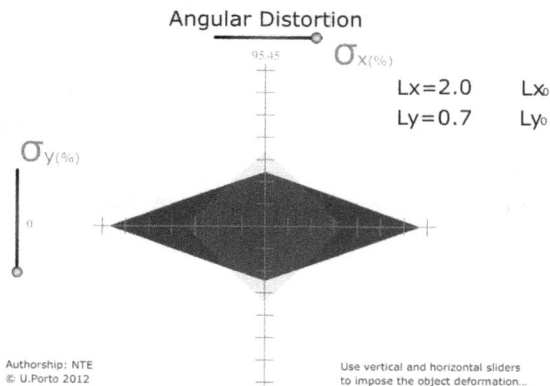

Biaxial_strain:

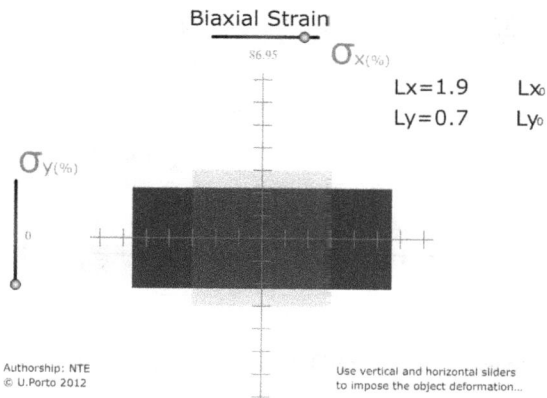

http://www.sensorsportal.com/DOWNLOADS/biaxial_strain.html

Uniaxial_strain:

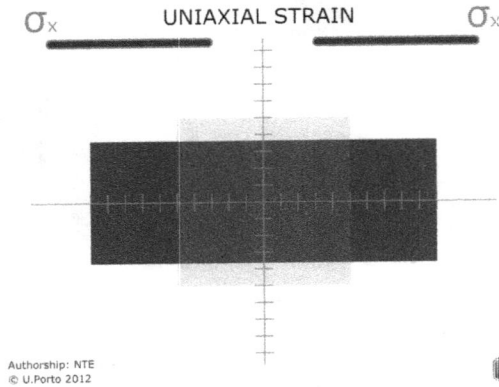

http://www.sensorsportal.com/DOWNLOADS/uniaxial_strain.html

Chapter 3:

Lead wire effects (Anim_condutores_ingles):

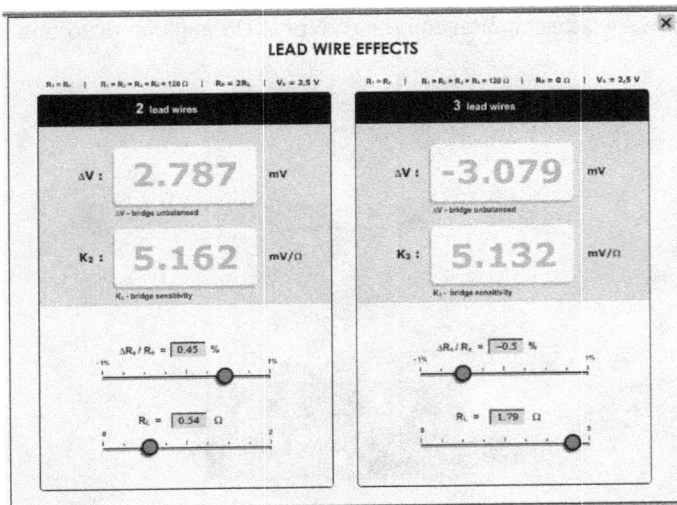

http://www.sensorsportal.com/DOWNLOADS/anim_condutores_ingles.swf

98

Compensation of temperature effects (Anim_temperatura_ingles):

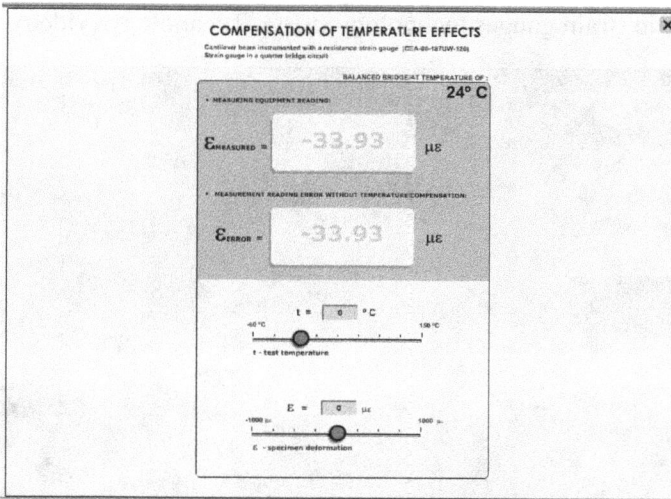

http://www.sensorsportal.com/DOWNLOADS/anim_temperatura-ingles.swf

Measurement of small resistance variations: potentiometer circuit/ Wheatstone bridge circuit (Anim_wheatstone_ingles):

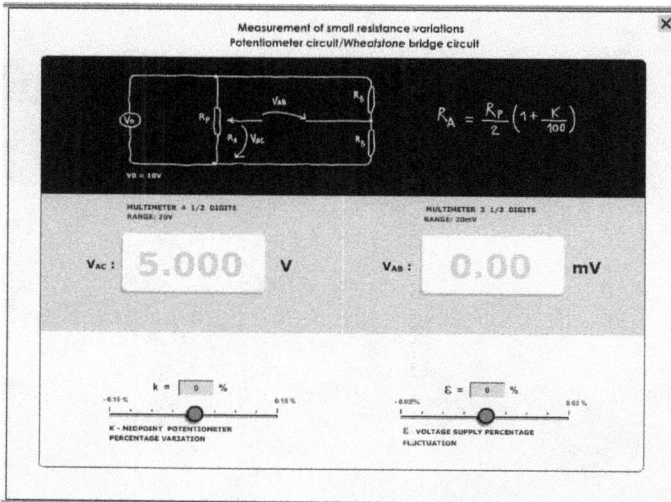

http://www.sensorsportal.com/DOWNLOADS/anim_wheatstone_ingles.swf

Chapter 4:

Fibre optic strain gauges technology (FBG_Technology (video)):

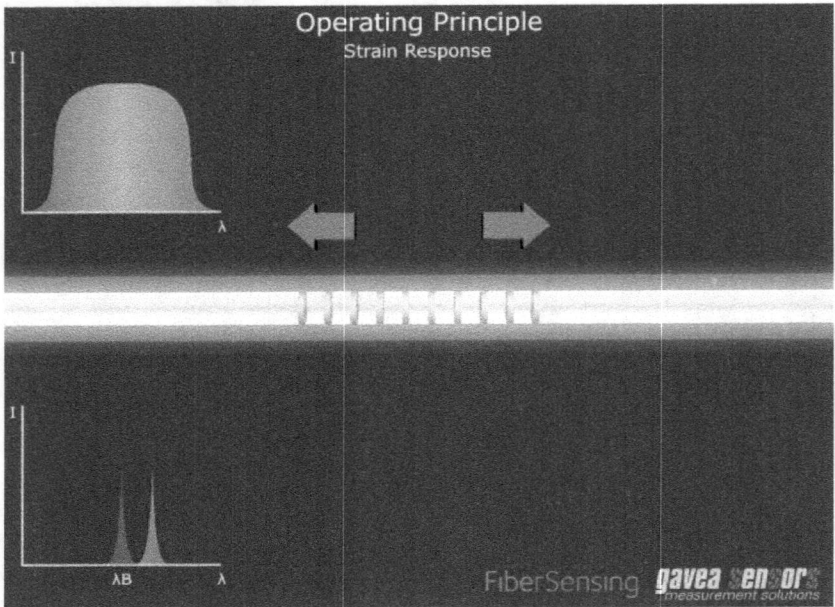

http://www.sensorsportal.com/DOWNLOADS/FBG_Technology.avi

Index

3-wire method, 46

A

active strain gauge, 53, 54

angular distortion, 18

apparent strain, 48

apparent strain of thermal origin, 50

B

biaxial stress state, 21

C

combined standard uncertainty, 82

combined uncertainty, 80

compressive deformation, 15

D

dummy strain gauge, 53, 54

E

elasticity modulus. *See* Young's modulus

elongation, 15

expanded uncertainty, 81

F

fibre Bragg gratings, 71

fibre optic strain gauges, 28, 65

G

Gauge Factor, 32

H

Hooke's law, 15

I

influence quantity, 81

interferometric strain sensing, 67

L

lead wire effects, 45

M

Mach-Zehnder configuration, 67

mechanical strain gauge, 27

metallic materials, 31

N

normal stresses, 17

O

optical fibre sensors, 65

P

plane stress state, 23

Poisson's ratio, 16